mvl

INTERNATIONAL
WILDLIFE
ENCYCLOPEDIA

THIRD EDITION

Marshall Cavendish Corporation
99 White Plains Road
Tarrytown, New York 10591–9001

Website: www.marshallcavendish.com

Library of Congress Cataloging-in-Publication Data

Burton, Maurice, 1898-
    International wildlife encyclopedia / [Maurice Burton, Robert Burton] .-- 3rd ed.
        p. cm.
    Includes bibliographical references (p.).
    Contents: v. 1. Aardvark - barnacle goose -- v. 2. Barn owl - brow-antlered deer -- v. 3. Brown bear - cheetah -- v. 4. Chickaree - crabs -- v. 5. Crab spider - ducks and geese -- v. 6. Dugong - flounder -- v. 7. Flowerpecker - golden mole -- v. 8. Golden oriole - hartebeest -- v. 9. Harvesting ant - jackal -- v. 10. Jackdaw - lemur -- v. 11. Leopard - marten -- v. 12. Martial eagle - needlefish -- v. 13. Newt - paradise fish -- v. 14. Paradoxical frog - poorwill -- v. 15. Porbeagle - rice rat -- v. 16. Rifleman - sea slug -- v. 17. Sea snake - sole -- v. 18. Solenodon - swan -- v. 19. Sweetfish - tree snake -- v. 20. Tree squirrel - water spider -- v. 21. Water vole - zorille -- v. 22. Index volume.
    ISBN 0-7614-7266-5 (set) -- ISBN 0-7614-7267-3 (v. 1) -- ISBN 0-7614-7268-1 (v. 2) -- ISBN 0-7614-7269-X (v. 3) -- ISBN 0-7614-7270-3 (v. 4) -- ISBN 0-7614-7271-1 (v. 5) -- ISBN 0-7614-7272-X (v. 6) -- ISBN 0-7614-7273-8 (v. 7) -- ISBN 0-7614-7274-6 (v. 8) -- ISBN 0-7614-7275-4 (v. 9) -- ISBN 0-7614-7276-2 (v. 10) -- ISBN 0-7614-7277-0 (v. 11) -- ISBN 0-7614-7278-9 (v. 12) -- ISBN 0-7614-7279-7 (v. 13) -- ISBN 0-7614-7280-0 (v. 14) -- ISBN 0-7614-7281-9 (v. 15) -- ISBN 0-7614-7282-7 (v. 16) -- ISBN 0-7614-7283-5 (v. 17) -- ISBN 0-7614-7284-3 (v. 18) -- ISBN 0-7614-7285-1 (v. 19) -- ISBN 0-7614-7286-X (v. 20) -- ISBN 0-7614-7287-8 (v. 21) -- ISBN 0-7614-7288-6 (v. 22)
    1. Zoology -- Dictionaries.   I. Burton, Robert, 1941-   . II. Title.

    QL9 .B796 2002
    590'.3--dc21
                                                                        2001017458

Printed in Malaysia
Bound in the United States of America

07 06 05 04 03 02 01   8 7 6 5 4 3 2 1

Brown Partworks
Project editor: Ben Hoare
Associate editors: Lesley Campbell-Wright, Rob Dimery, Robert Houston, Jane Lanigan, Sally McFall, Chris Marshall, Paul Thompson, Matthew D. S. Turner
Managing editor: Tim Cooke
Designer: Paul Griffin
Picture researchers: Brenda Clynch, Becky Cox
Illustrators: Ian Lycett, Catherine Ward
Indexer: Kay Ollerenshaw

Marshall Cavendish Corporation
Editorial director: Paul Bernabeo

Authors and Consultants

Dr. Roger Avery, BSc, PhD (University of Bristol)

Rob Cave, BA (University of Plymouth)

Fergus Collins, BA (University of Liverpool)

Dr. Julia J. Day, BSc (University of Bristol), PhD (University of London)

Tom Day, BA, MA (University of Cambridge), MSc (University of Southampton)

Bridget Giles, BA (University of London)

Leon Gray, BSc (University of London)

Tim Harris, BSc (University of Reading)

Richard Hoey, BSc, MPhil (University of Manchester), MSc (University of London)

Dr. Terry J. Holt, BSc, PhD (University of Liverpool)

Dr. Robert D. Houston, BA, MA (University of Oxford), PhD (University of Bristol)

Steve Hurley, BSc (University of London), MRes (University of York)

Tom Jackson, BSc (University of Bristol)

E. Vicky Jenkins, BSc (University of Edinburgh), MSc (University of Aberdeen)

Dr. Jamie McDonald, BSc (University of York), PhD (University of Birmingham)

Dr. Robbie A. McDonald, BSc (University of St. Andrews), PhD (University of Bristol)

Dr. James W. R. Martin, BSc (University of Leeds), PhD (University of Bristol)

Dr. Tabetha Newman, BSc, PhD (University of Bristol)

Dr. J. Pimenta, BSc (University of London), PhD (University of Bristol)

Dr. Kieren Pitts, BSc, MSc (University of Exeter), PhD (University of Bristol)

Dr. Stephen J. Rossiter, BSc (University of Sussex), PhD (University of Bristol)

Dr. Sugoto Roy, PhD (University of Bristol)

Dr. Adrian Seymour, BSc, PhD (University of Bristol)

Dr. Salma H. A. Shalla, BSc, MSc, PhD (Suez Canal University, Egypt)

Dr. S. Stefanni, PhD (University of Bristol)

Steve Swaby, BA (University of Exeter)

Matthew D. S. Turner, BA (University of Loughborough), FZSL (Fellow of the Zoological Society of London)

Alastair Ward, BSc (University of Glasgow), MRes (University of York)

Dr. Michael J. Weedon, BSc, MSc, PhD (University of Bristol)

Alwyne Wheeler, former Head of the Fish Section, Natural History Museum, London

Picture Credits
Ardea London: Eric Dragesco 422, Jean-Paul Ferrero 421, J.M. Labat 293, J.L. Mason 416, Stefan Meyers 294; **Neil Bowman:** 309, 320, 324, 384, 399; **Bruce Coleman:** Franco Banfi 390, Jen and Des Bartlett 329, Erwin and Peggy Bauer 295, 300, 372, 386, 427, Erik Bjurstrom 389, Mr. J. Brackenbury 318, Jane Burton 314, 330, 366, 396, John Cancalosi 301, 353, 426, Alain Compost 306, 404, Bruce Coleman Ltd 303, 304, Dr. P. Evans 374, M.P.L. Fogden 313, 356, Christer Fredriksson 332, C.B. and D.W. Frith 398, Paul Van Gaalen 327, Bob Glover 367, Dennis Green 414, HPH Photography 338, 403, Johnny Johnson 391, 392, Janos Jurka 365, Stephen C. Kaufman 385, 394, P. Kaya 418, 420, Stephen J. Krasemann 387, 388, Felix Labhardt 352, Harald Lange 305, Wayne Lankinen 319, Werner Layer 322, 361, Robert Maier 344, 395, Luiz Claudio Marigo 359, 360, 382, 383, George McCarthy 400, Joe McDonald 312, 323, 371, Michael McKavett 331, 415, Tero Niemi 292, 308, 376, 378, Charlie Ott 424, Mary Plage 343, Dr. Eckart Pott 307, 370, Andrew Purcell 299, 328, 342, Hans Reinhard 298, 347, 351, 364, 409, 412, 425, Dr. Frieder Sauer 327, John Shaw 428, Kim Taylor 296, 297, 316, 345, 419, Norman Tomalin 373, Uwe Walz 358, Jorg & Petra Wegner 368, 375, Staffan Widstrand 393, Wild-Type Productions 321, Rod Williams 333, 406, Gunter Ziesler 325, 405; **Corbis:** Richard Cummins 402, Frank Blackburn/Ecoscene 310, Dan Guravich 346, Pam Gardner/Frank Lane 340, Wolfgang Kaehler 379, 411, S.F. Morton Beebe 397; **NHPA:** A.N.T. 341, 349, 408, Tom Ang 363, Daryl Balfour 337, Anthony Bannister 417, Stephen Dalton 355, Manfred Danegger 311, John B. Free 317, Melvin Grey 354, Martin Harvey 413, Paal Hermansen 377, Daniel Heuclin 348, 423, David Middleton 380, Christophe Ratier 335, 336, Jany Sauvanet 381, Dave Watts 302, Norbert Wu 407. **Artwork:** Catherine Ward 326.

Contents

BROWN BEAR

The brown bear was formerly widespread throughout Europe but is today found only in remote forests on the continent. This bear was photographed in Finland.

THE BROWN BEAR IS a heavily built member of the order Carnivora. It is practically tailless, with broad, flat feet, each of which has five toes armed with nonretractile claws. The brown bear's eyesight is relatively poor, but its smell and hearing are both acute. The snout is well developed, with a wet nose, or rhinarium, which heightens its sense of smell. The usual coat color is a shade of brown, but bears found in central Asia are reddish and those that occur in western China and Tibet have blackish brown hairs frosted with slate gray.

It is a matter of opinion whether there are several species of brown bear distributed over Europe, Asia and North America, or whether there is only one. Half a dozen species, as well as many subspecies, have been recognized by various authorities in the past. This is now accepted as being too many, given that all brown bears have a similar habitat, behavior and life history, and that their skeletons and general anatomy are not significantly different. Today the consensus is that the grizzly bear and the other brown bears of North America, such as the giant Kodiak bear of Alaska and northern Canada, belong to the same species as the smaller brown bears found in Europe, the former Soviet Union, the Middle East and Central Asia.

Regional variation

The historical list of species and subspecies of brown bear was based mainly on differences in coat color and size. Body coloration is variable in many mammals, however, and so size was considered more important. The Kenai and Kodiak bears can weigh as much as 1,700 pounds (770 kg), whereas Siberian brown bears weigh 330–550 pounds (150–250 kg) and the grizzly

BROWN BEAR

CLASS	**Mammalia**
ORDER	**Carnivora**
FAMILY	**Ursidae**
GENUS AND SPECIES	***Ursus arctos***

ALTERNATIVE NAMES

Grizzly bear (*U. a. horribilis*); Kodiak bear (*U. a. middendorffi*); European brown bear (*U. a. arctos*); Tibetan brown bear (*U. a. pruinosus*)

WEIGHT

Up to 1,700 lb. (770 kg)

LENGTH

Head and body: 5⅔–9⅕ ft. (1.7–2.8 m); shoulder height: 3–5 ft. (0.9–1.5 m); tail: 2–8 in. (6–21 cm)

DISTINCTIVE FEATURES

Prominent shoulder hump; long fur and claws

DIET

Mainly grasses, roots, berries, fish, insects, honey, mammals and carrion

BREEDING

Age at first breeding: 4–6 years or over; breeding season: May–July; number of young: usually 2; gestation period: 180–270 days; breeding interval: 2–4 years

LIFE SPAN

Up to 50 years in captivity

HABITAT

Tundra, alpine meadows, coasts, coniferous forest (Siberia), montane woodland (Europe)

DISTRIBUTION

Western North America; Scandinavia east to easternmost Asia and Japan; mountains of Iberia, central Europe and Balkans

STATUS

Numbers vary greatly according to area; population: 100,000, Eurasia; 50,000, Canada and Alaska; 1,000, rest of U.S.

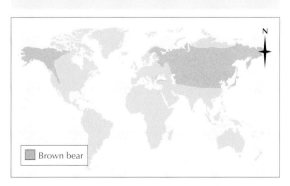

Brown bear

bears of Yellowstone National Park weigh only 225–715 pounds (100–325 kg). In 1904 J. G. Millais, naturalist and author, wrote that "no terrestrial mammal varies so greatly, both in size and pelage [coat coloration], as this animal."

Although cubs and young bears can climb trees with relative ease, adults do not often attempt such a feat.

Loners

Brown bears live in forests, the northern tundra and wild, mountainous country. They wander singly or in family parties over home ranges that have an average radius of 20 miles (32 km), although individuals may stray beyond their usual range. Brown bears normally walk on all fours, at times standing erect and shuffling for a few paces. Young bears climb trees well,

The salmon streams in Alaska, northern British Columbia and Kamchatka (Siberia) offer an abundant supply of fish, which are high in fat and easy to catch.

although slowly and deliberately, but adults rarely do so. Bears feed intensively during the summer and fall, storing fat in their bodies. This enables them to sleep for most of the winter in dens, located among rocks and inside hollow trees, or dug into hillsides.

Brown bears are not normally aggressive to humans except under extreme provocation, when they are injured, or when a person comes between a female and her cub. The strength of brown bears is demonstrated by the fact that a grizzly bear can fell an American bison larger than itself with one blow of its paw, breaking its neck. The bear is also strong enough to drag away the carcass.

Omnivorous carnivores

Despite the fact that brown bears are classified in the order Carnivora, or flesh-eaters, their diet varies greatly with the individual. Some are wholly vegetarian, others wholly flesh-eating, but most eat both plant and animal foods.

The diet of brown bears typically consists of berries, fruits, roots, insects and other small animals, honey and the grubs of wild bees. Fish may be taken, flipped from the shallows onto the bank with the forepaws, or seized in the water, as is the case in the salmon runs of western North America. Sometimes young deer are killed and eaten, and larger mammals, such as bison and moose, are also taken. Occasionally brown bears will kill livestock and even the smaller American black bears.

Tiny cubs

Brown bears lead an existence similar to that of black bears. Mating takes place in June, but the fertilized egg is not implanted into the wall of the uterus until October or November, a bodily process known as delayed implantation. The overall gestation period is therefore 180–266 days. As a result the cubs, which normally number two, are born in January or February, while the mother is still in her winter sleep. Each cub weighs 1–1½ pounds (450–680 g) at birth and is 8 inches (20 cm) long, almost hairless, blind and toothless.

Brown bear cubs are tiny compared with the size of adult females, and an ancient belief was that they were born shapeless and that the mother licked them into shape, hence the well-known saying. The reference here is to the licking the mother gives each cub after birth to clean them of the birth fluids, as in most true mammals. The mother rouses herself from her sleep to do so. A brown bear cub weighs 54 pounds (24 kg) when a year old, and will stay with its mother until at least that age. Females usually breed at 4–6 years of age, and individual bears have been known to live for up to 50 years.

Shrinking range

In the Old World brown bears once ranged in considerable numbers from Britain east to Japan, and as far south as the Mediterranean, the Himalayas and northeastern Africa. By the 11th century the last bear had been killed in Britain and today, in Europe, the surviving descendants are largely confined to inaccessible forests in the Pyrenees, the Swiss Alps, the Carpathian mountains, the Balkans and areas of Scandinavia. Brown bears are more numerous in some regions of Asia, particularly parts of the former Soviet Union, but even there numbers are steadily declining. The North African brown bear population became extinct in the 19th century.

It is evident that persecution, particularly hunting, radically reduces the maximum sizes reached in a species, especially when the hunters' ambition is to collect record trophies. Brown bears were killed for their flesh and their fat, and because their tempers were unpredictable. Above all, however, their shaggy coats were coveted prizes for hunters. Today C.I.T.E.S. (Convention on International Trade in Endangered Species of Wild Flora and Fauna) lists all U.S. populations of *Ursus arctos* as threatened, and the Mexican brown bear (*U. a. nelsonii*), Tibetan brown bear (*U. a. pruinosus*) and Italian populations of *U. arctos* as endangered. The I.U.C.N. (World Conservation Union) classifies *U. a. nelsonii* as extinct.

The cave bear

Bones of the now-extinct cave bear, *U. spelaeus*, have been found in caves throughout western Europe. This species denned in caves during the winter and females may also have given birth in caves. The cave bear was primarily herbivorous, although it would occasionally have eaten meat; it had a large head and massive canine teeth. The cave bear also possessed powerful front legs and paws and large, broad claws, which suggests that it dug for much of its food. The root-beds of plants buried in deep glacial silt must have formed a large part of its diet.

When the large glaciers that covered much of western Europe melted, the cave bear began to reduce in size. It may have become smaller because of a decrease in the amount of root-beds due to low glacial silt production. The growing populations of brown bears in Europe during the late Pleistocene (the last ice age) may also have competed with cave bears for food resources, driving the species to eventual extinction.

The rich food supplies and favorable climate around Kodiak Island, Alaska, have enabled the Kodiak brown bear to attain the largest size and weight of the species.

BROWN BUTTERFLIES

The meadow brown (female, above) is one of the most common European butterflies.

MEMBERS OF THE BUTTERFLY subfamily Nymphalidae are found worldwide, from the Tropics to the Arctic Circle and from sea level to high mountains. There are about 1,500 species in all. Due to the rather somber coloration of many species, they are known as brown butterflies, a name frequently shortened to the browns.

A particularly common and well-studied Old World species of brown butterfly is the meadow brown, *Maniola jurtina*. It is found in Europe and North Africa in a wide range of grassy areas, including meadows, rough pasture, hedgerows, woodland clearings and disturbed ground, such as at roadsides. Where this type of habitat is allowed to persist, the meadow brown may be abundant. There are numerous local variations in its coloration and many closely related species throughout its range.

Male meadow browns are relatively dull colored, the females being larger and brighter with an area of dull orange on the forewings. This orange area is much less extensive in males. Like a number of other butterflies the male has a patch of scented scales, known as androconia, on each forewing. These scales play an important role during courtship. Their scent has been described as resembling that of an old cigar box.

Unobtrusive caterpillars

In midsummer meadow browns are on the wing, ready for mating. The male approaches a female and the scented scales on his wings make her responsive. After mating the female lays a batch of whitish-green eggs shaped like minute barrels with vertically ribbed surfaces. They are laid on the blades of various grasses, which are the food plant of the larvae, or caterpillars. These are green with short white hairs and have a dark line along the back and a lighter one on each side. There are two short points at the hind end, a feature almost universal among larvae of the subfamily Satyrinae. Meadow brown larvae are seldom seen because they feed mainly at night, hiding at the base of grass stems by day. They have a long life, often from early August to the following April, although this varies according to environmental factors.

MEADOW BROWN

CLASS	**Insecta**
ORDER	**Lepidoptera**
FAMILY	**Nymphalidae**
SUBFAMILY	**Satyrinae**
GENUS AND SPECIES	***Maniola jurtina***

LENGTH
**Adult wingspan: 1½–2¼ in. (4–5.5 cm).
Larva length: up to 1 in. (2.5 cm).**

DISTINCTIVE FEATURES
Adult: wings mainly brown and orange, with large black eyespot on forewings. Larva: green, with faint markings.

DIET
Adult: flower nectar. Larva: grasses.

BREEDING
Number of eggs: several hundred; hatching period: about 30 days; breeding interval: 1 year

HABITAT
Meadows, roadsides, woodland clearings and other grassy areas, from sea level to 5,000 ft. (1,525 m)

DISTRIBUTION
Europe east to the Ural Mountains and south to parts of North Africa

STATUS
Very common

Meadow brown

Most, but not all, of the brown butterflies have subdued brown, gray and orange wings.

The name brown butterflies was applied by European entomologists unfamiliar with the browns of the Tropics and subtropics. The majority of species in Satyrinae certainly fit the name, although they may often have intricate patterns in brown and orange, but there are many brightly colored representatives in warmer regions. In South Africa, for example, there are several blue species of Satyrinae. A few satyrids are silver and most unusual in appearance. In the depths of the Amazon rain forest live several transparent satyrine butterflies, called clearwings.

Many butterflies alter their position when they land in bright sunshine in order to cast the smallest possible shadow and make themselves less conspicuous to predators. This can be seen in the European grayling butterfly, *Hipparchia semele*, which generally sits on exposed ground. On landing it often leans forward toward the sun's rays, reducing its shadow length. Body positioning such as this may also be connected with temperature regulation. Certain species of butterflies fold their wings when at rest to hide the more brightly colored upperwings, exposing only the drab underwings. In the Tropics some satyrine butterflies take such predator avoidance behavior a stage further by emerging only at dusk; these species are called evening browns.

Perfumed courtesy

The most complete study of courtship in the subfamily Satyrinae has been of the grayling. The male grayling chases the female through the air until she lands, whereupon he takes a position facing her and so close that her antennae overlap his head and the front of his thorax. The organs of smell are in the antennae. Once settled, the male begins to court the female, closing his wings in such a way that the female's antennae are caught between his forewings and pressed against his scent patches.

BROWN RAT

Highly adaptable to different conditions, the brown rat is found in virtually every terrestrial habitat except polar and desert regions.

THE BROWN RAT IS one of the world's most widespread and abundant mammals. It is considerably larger than its close relative, the ship, or black, rat, reaching up to 11 inches (28 cm) from nose to the base of the tail and weighing up to 1¾ pounds (790 g). In comparison with the ship rat, the brown rat has smaller ears and eyes, a proportionately shorter and fatter tail and shaggier fur. Its coat is usually brown or brownish gray but can vary greatly in color and for this reason the species is often referred to more generally as the common rat. The white rats used in laboratories and kept as pets are specially bred albino varieties of brown rats.

Brown rats are regarded by humans as a pest, inhabiting buildings and attacking crops both before harvest and when they are stored in warehouses. In the past brown rats followed foodstuffs aboard ships, which then carried the rats around the world. Today the species is found from Alaska south to the whaling stations of South Georgia island in the southern Atlantic and is widespread in Europe, Asia and Africa.

The origins of the brown rat are not known for certain. It is thought that the species spread across Europe from central Asia, and at one time it was suggested that brown rats arrived in Britain onboard ships coming from Norway at the beginning of the 18th century. For this reason the species was given the scientific name *Rattus norvegicus* and became known as the Norway rat.

Scientists now believe that brown rats arrived in northwestern Europe between 1728 and 1730 in trading ships from East Asia. At the same time the rats were spreading overland across Europe.

On the march

Swarms of rats on the march were often reported in the early 18th century although vigorous control campaigns have curbed the spread of the species. Even now, however, there are sporadic reports of rat migrations when local populations build up to plague proportions. When they have no migratory urge, brown rats keep to a fairly restricted home range in warehouses, farm buildings, corn ricks, garbage dumps, hedgerows and woodlands. They are often found along riverbanks and, being good swimmers and divers, they are sometimes mistaken for water voles.

The brown rat's home range, or territory, is inhabited by a family group that lives together; there is little fighting except by females defending their nests. Rats from other groups are not tolerated and there are border skirmishes where territories meet. If a male from an outside group is placed in a colony of captive rats from which it cannot escape, it is usually killed within hours.

Within the home range there is a system of runs, regularly used by the rats while they are foraging. These are often underground, but when above ground they are positioned to take advantage of natural cover, for the runs give protection from predators. Brown rats mark out their runs with scent marks of urine and learn the routes thoroughly, enabling them to speed along the runs when danger threatens. The rats also communicate by means of ultrasonic hearing and through a wide range of physical and behavioral postures and signals.

Destructive mammals

Brown rats feed on a diverse assortment of foods. Because of the versatility of their feeding habits and the enormous size attained by their populations, brown rats can be extremely destructive. Their main food is grain, though they are also a serious pest of sugar beet and other root crops.

Brown rats eat more flesh than most rodents and, when they live in meat stores, may eat little else. They consume a wide variety of animals; the remains of house mice, the skins turned inside out, are often found near rats' nests. Brown rats will attack poultry runs, often carrying eggs away to their burrows to feed on; they will also attack ducklings in the water. In addition, brown rats are major carriers of diseases,

BROWN RAT

CLASS	**Mammalia**
ORDER	**Rodentia**
FAMILY	**Muridae**
GENUS AND SPECIES	***Rattus norvegicus***

ALTERNATIVE NAMES
Common rat; Norway rat

WEIGHT
Up to 1¾ lb. (790 g), average 1¼ lb. (500 g)

LENGTH
**Head and body: 11 in. (28 cm);
tail: 4¾–5½ in. (12–14 cm)**

DISTINCTIVE FEATURES
**Long, scaly tail slightly shorter than body;
pointed muzzle; large feet; smaller eyes and
ears than ship (black) rat**

DIET
**Virtually anything digestible; grain and root
crops widely eaten**

BREEDING
**Age at first breeding: about 11 weeks;
breeding season: all year; number of young:
usually 7 or 8; gestation period: 21–24 days;
breeding interval: commonly 3 to 5 litters
per year**

LIFE SPAN
Up to 1 year; up to 5 years in captivity

HABITAT
**Nearly all terrestrial habitats except deserts,
tundra and polar ice; well adapted to
artificial habitats**

DISTRIBUTION
**Found worldwide wherever humans live
in concentrated areas; absent only from
northernmost regions of Northern
Hemisphere and parts of South America,
Africa and Central Asia**

STATUS
Globally abundant; populations in billions

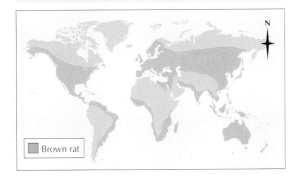

Brown rat

including Weil's disease, typhus and salmonella. The carriers of the flea that transmits bubonic plague are black rats, not the common brown rats.

Prolific breeders

There is little or no courtship before two brown rats mate and a bond never forms between the pair. The female builds a spherical nest from any material available, and the litter is born there after a gestation of 21–24 days. The young, up to to 11 in number depending on the body weight of the mother, are born naked and helpless. Their eyes and ears open a few days later and they leave the nest when they are 3 weeks old.

When the brown rat's home range is in an environment that alters little during the year, such as a wheat barn, breeding takes place all the time. This does not mean that females breed year-round; each produces only three to five litters a year, but there are always some females in breeding condition. In fields and hedgerows there are one or two annual peaks of breeding.

A female brown rat begins breeding when she is about 11 weeks old and can then produce up to 50 young a year. This figure is never reached in practice because mortality, especially of young rats, becomes very high when the population is large. As many as 99 percent of the animals in a population will die before reaching adulthood, and most adults do not live for more than a year. This very rapid rate of reproduction and turnover means that brown rats are very difficult to control.

Owls, weasels and foxes all hunt brown rats, although even a weasel may have difficulty killing a healthy, full-grown rat. Around farmyards domestic cats kill many rats. Although cats alone cannot reduce rat populations, if rat poison is also employed the cats can prevent the rats from regaining their former numbers.

Brown rats readily feed on grain and crops, both before harvest and after storage, and are regarded as a major agricultural pest.

BRUSH TAIL POSSUM

Brush tail possums are Australasia's most common marsupials.

Brush tail possums are largely vegetarian animals and feed mainly on buds, shoots, flowers, leaves, fruits and seeds. The Australian mistletoe in particular is a favorite food. The possums will also eat insects and are reputed to take eggs and nestling birds.

Daytime shelters
Brush tail possums active in the open during the day are likely to fall prey to eagles. Those without adequate sleeping shelter are equally vulnerable to attack; as a result, the number of brush tail possums in a given area is often determined by the availability of hollow trees and burrows. Dingoes will also hunt possums, and will tear bark from the base of a hollow tree if they believe that a possum is hidden inside.

The main predator of the brush tail possum is a species of monitor lizard, *Varanus varius*, known as the goanna. If a brush tail possum sees a goanna crawling up the tree in which it is situated, it will cry out in alarm. The Aborigine people of Australia, who cook and eat brush tail possums, will imitate the goanna's scratching on the bark of a tree to encourage any possums within to cry out and reveal themselves.

Young reared in mother's pouch
Mating takes place throughout the year, with peaks in the spring and the fall. The young are usually born singly, though occasionally twins or even triplets are born at the same time. Young possums leave the pouch 3–4 months after birth, becoming independent of the mother in a further 3 months. At birth a young possum is ½ inch (1.5 cm) in length and weighs less than $\frac{1}{10}$ ounce (2 g), compared with the mother's weight of about 10 pounds (4.5 kg). The young of *T. vulpecula* and *T. arnhemensis* attain sexual maturity some 9–12 months after birth. However, the young of the species *T. caninus* are not sexually mature until 24–36 months of age.

As with other baby marsupials, the newly born brush tail possum makes its way unaided from the birth canal to the pouch through the mother's fur. It progresses by an overarm action of the front legs, which at birth are longer than the hind legs, the paws being armed with strong claws. The paws can be flexed to grasp the mother's fur. It takes the newly born possum over 5 minutes to travel the 2½ inches (6.5 cm) to the pouch. Once inside, the young possum seizes one of the mother's two teats in its mouth and does not abandon it for some weeks. At the end of this period, the possum begins to take on a

There are three species of brush tail possum: *Trichosurus vulpecula*, the most common, *T. arnhemensis* and *T. caninus*. These species are the most abundant and widely distributed of all Australasian marsupials. They flourish in New Zealand, where they were introduced in 1858. Brush tail possums are up to 23 inches (58 cm) long, similar to the size of the red fox, and have a foxlike head with large ears and a pointed snout. Their most distinctive feature, however, is the tail; this is prehensile at the tip, and there is a naked patch on the underside. The fur of brush tail possums is thick, woolly and variable in color, ranging from silver-gray to dark brown or black.

Adaptable marsupials
Brush tail possums usually live in trees but will also inhabit low bush and treeless areas, where they take over rabbit burrows. They are also frequently found living in the roof-spaces of houses, even those that are located in the suburbs of large towns.

All three species of brush tail possum appear to be indifferent to humans. They may allow themselves to be stroked, though they will object to being picked up. This passive behavior contrasts with the noise created when these animals quarrel with each other; during these confrontations the possums hiss and grunt, emitting a loud cry that ends in a piercing screech.

BRUSH TAIL POSSUMS

CLASS	**Mammalia**
ORDER	**Marsupialia**
FAMILY	**Phalangeridae**

GENUS AND SPECIES **Trichosurus vulpecula, T. arnhemensis, T. caninus**

ALTERNATIVE NAMES
Vulpine possum; foxlike possum

WEIGHT
3–11 lb. (1.3–5 kg)

LENGTH
Head and body: 12½–23 in. (32–58 cm); tail: 9½–14 in. (24–35 cm)

DISTINCTIVE FEATURES
Thick, wool-like fur; large ears; pointed snout; naked patch on underside of prehensile tail

DIET
Shoots, leaves, flowers, fruits, seeds and insects; occasionally eggs and young birds

BREEDING
Age at first breeding: 9–12 months (T. vulpecula, T. arnhemensis), 24–36 months (T. caninus); breeding season: all year, with peaks in spring and fall; number of young: usually 1; gestation period: 16–18 days; breeding interval: usually 1 year, occasionally 2 litters per year

LIFE SPAN
Up to 13 years (T. vulpecula); up to 17 years (T. caninus)

HABITAT
Forest and suburban gardens; also semi-deserts and other treeless areas (T. caninus)

DISTRIBUTION
Australia; introduced to New Zealand

STATUS
Common

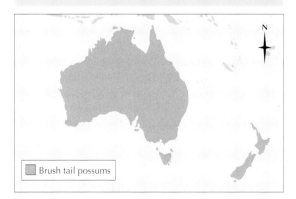

Brush tail possums

more adult appearance. If the young possum fails to reach the pouch, or is otherwise lost, the mother will give birth to another shortly.

Secrets of survival

The population of brush tail possums in New Zealand reached 25 million in the 1960s. Humans are the only predators of possums in this area of the world. About a million possums a year are trapped for their fur, and another million are destroyed as pests under a bounty system. Despite these losses and the relatively low birth rate of the species, there is an annual population increase of some 25 percent.

In Australia brush tail possums are regarded as a pest in orchards and on farms. In the past the population of the animals underwent a decline because they were killed for their valuable fur, which was exported and sold under the names of beaver, skunk and Adelaide chinchilla. Today there is a more sympathetic attitude toward brush tail possums, based partly on the increasing realization of the need to conserve Australia's unique fauna. It has also been realized that, by feeding on mistletoe, brush tail possums check the spread of this parasitic plant, benefiting indigenous trees. Conservation of the species also indirectly helps the Australian honey industry. The economic value of brush tail possums in reducing the prejudicial effect of mistletoe on flowering gum trees, which produce nectar on which honeybees feed, has been proven by direct experiment.

Although hunted for its fur and as a pest, the brush tail possum is not threatened, primarily because of the species' ability to survive in different habitats. However, the need for local control of the brush tail possum is now recognized.

The newly born brush tail possum remains within its mother's pouch for several weeks. By the end of this period it has grown fur and started to look like an adult.

301

BUDGERIGAR

Many different varieties of budgerigar have been selectively bred worldwide. In its wild form the species is primarily green and yellow.

THE BUDGERIGAR IS THE commonest member of the parrot family in Australia; flocks numbering thousands gather in the arid parts of the continent. At about 7 inches (18 cm) long the budgerigar is the smallest species of parakeet in the subfamily Psittacinae. It is mostly yellow and green, the sexes having very similar coloring. The Aboriginal name for the budgerigar is *betcherrygah*, meaning "good food," and zoologists believe that Aborigines have been eating the birds for a long time without affecting their numbers. An alternative Australian spelling of the name is budgerygah.

The wild budgerigar is grass green on the back and underparts with bright yellow on the head and blue on the tail, which is long and tapering, as in many other species of parakeet.

The upperparts are barred and scalloped with black and yellow; there is an intense blue patch on each cheek and three black spots on each side of the throat. The male differs from the female in that the base of the bill is blue instead of brownish.

Budgerigars are among the most familiar of all cagebirds and breed freely in captivity. Large numbers have been exported from Australia, and many more have been captive-bred worldwide. Selective breeding has produced many color varieties, including various shades of green, blue, yellow, gray, olive and white.

Nomadic flocks

Budgerigars are gregarious and highly nomadic birds which travel in large flocks. They have no fixed territories outside the breeding season, and

BUDGERIGAR

CLASS	**Aves**
ORDER	**Psittaciformes**
FAMILY	**Psittacidae**
GENUS AND SPECIES	***Melopsittacus undulatus***

ALTERNATIVE NAMES
Budgerygah; betcherrygah; canary parrot; lovebird; flightbird

WEIGHT
Usually about 1½ oz. (40 g)

LENGTH
Head to tail: 6¾–8 in. (17–20 cm)

DISTINCTIVE FEATURES
Slender body; mainly bright green with yellow throat and forehead and barring on upperparts; long, tapering blue tail

DIET
Mainly seeds and grain; some fruits and other plant material

BREEDING
Age at first breeding: 3 months; breeding season: March–September (north), August–January (south); number of eggs: usually 4 to 6; incubation period: 18–20 days; fledging period: 30–36 days; breeding interval: 1 or 2 broods per year

LIFE SPAN
Over 10 years in captivity

HABITAT
Arid and semiarid woodlands and grasslands

DISTRIBUTION
Inland Australia

STATUS
Abundant

Budgerigar

move instead from one feeding ground to another. The original diet of budgerigars was the seeds of various wild grasses, supplemented with other seeds and fruits. However, the spread of agriculture in Australia has resulted in cereal crops developing into a major source of food for the birds.

The typical budgerigar call is a churring warble interspersed at times with whistles and screams. This warble is continuous when flocks drink at waterholes and streams together. In captivity budgerigars readily mimic other sounds, including human speech. When they are alarmed, budgerigars chatter.

Hole-nesters

Breeding usually takes place in mallee scrub, an Australian eucalypt tree that resembles silver birch. The tree has several trunks, which spring from a common base. Where the stems meet, moisture is trapped, causing the wood to rot, and it is in these rotting crevices that budgerigars make their nests, chipping away with their bills to make holes 6–12 inches (15–30 cm) deep and about 2 inches (5 cm) wide. The birds will also

Flocks of budgerigars congregate during the morning and afternoon to drink. Each flock may consist of hundreds or even thousands of birds.

Budgerigars are highly nomadic. They move periodically from one feeding ground to another and have no fixed territories outside their breeding season.

nest in hollow cavities in tree stumps or in fence posts. Budgerigars use no other nesting material apart from chips of rotten wood and wood dust lying in the holes. Several pairs of budgerigars may nest in close proximity, some sharing the same hollow.

Responsive to the male's song

Experiments have shown that in many bird species the proximity of a male, or even his courtship, is insufficient to stimulate egg-laying in the female. This is true of the female budgerigar, which requires total darkness for 24 hours each day, or a nest box that simulates the conditions inside a nest in the wild. When researchers gave unpaired females suitable nest boxes and allowed them to hear the song of male budgerigars, the females laid within 18 days. Females kept under identical conditions but not allowed to hear a male's voice failed to lay after a month.

The male budgerigar courts the female by nudging her bill, bobbing his head toward her and offering gifts of food. He makes a number of different sounds including warbling, which occurs when testosterone is released into the blood, signaling that he is in breeding condition.

Perhaps the most conclusive results came from those tests in which tape recordings were used. A female within a nest box and kept either entirely in the dark or exposed to light for only brief periods would lay eggs on being allowed to hear a tape recording of a male's warble. If allowed to hear this song for a total of 6 hours a day the female was ready to lay in a shorter period of time than if presented with the warble for three hours a day. However, breeding is also influenced by other factors, including rainfall.

Female budgerigars are able to reproduce after 3 months of age; they lay up to nine eggs at a time and there may be two broods a year. The eggs are spherical and white, and are incubated for 18–20 days by the hen alone, which is fed by the male during this time. The nestlings open their eyes about 8 days after hatching and remain in the nest for about 4–5 weeks; they are tended by the parents until fully fledged.

BUFFALO

THERE ARE 13 SPECIES of buffalo, all of which are large animals. Domesticated Indian, or water, buffalo can weigh over a ton, with some males reaching as much as 2,600 pounds (1,200 kg). American and European bison may reach 1,800 pounds (800 kg). Perhaps surprisingly considering its fearsome reputation, the Cape buffalo of Africa weighs less than these species, although large males can still tip the scales at nearly 1,500 pounds (700 kg).

Most of the other buffalo are less well known, and all but two are threatened or endangered. Among them are such heavyweights as the Himalayan yak, which has an exceptionally long, shaggy coat, and the gaur of Southeast Asia, which is sometimes known as the Indian bison. The tamarau and the lowland and mountain anoas are relatively small for buffalo; some grow to no more than 650 pounds (300 kg). They inhabit the humid rain forests of Indonesia and the Philippines, where they cling to survival in much reduced ranges.

Raging bulls

A large mass means that buffalo are among the most dangerous of all mammals in terms of human fatalities. Cape buffalo rank as one of Africa's so-called big five quarry species, valued and feared in equal measure by game hunters and safari-goers. They cause as many fatalities as other, more notorious human-killers, including lions, elephants and hippos. The few bison surviving in North America's national parks are sometimes taunted by reckless tourists and as a result are responsible for stampeding and trampling small numbers of unlucky and foolhardy spectators every year. All wild buffalo have a variety of dominance and threat displays that are most aggressive when performed by bulls (mature males). Old bulls often bear the scars of many past battles with rivals.

The yak belongs to the genus Bos, *or true oxen. It is a high-altitude resident of the Tibetan plateau in Central Asia.*

CLASSIFICATION

CLASS	Mammalia
ORDER	Artiodactyla
FAMILY	Bovidae
SUBFAMILY	Bovinae
TRIBE	Bovini
NUMBER OF SPECIES	Genus *Bison*: 2 Genus *Syncerus*: 1 Genus *Bubalus*: 4 Genus *Bos*: 6

Grazing specialists

Buffalo include all the various breeds of domestic cattle, and make up the tribe Bovini. This group belongs to the subfamily Bovinae, which in turn is contained within the large family Bovidae. Other members of the Bovidae include antelopes, sheep, goats and the pronghorn antelope. Buffalo are also related to deer, camels, hippos and giraffes. In common with all of these grazing mammals, buffalo have a large multichambered stomach. One chamber, the rumen, harbors bacteria that digest the tough cell walls of grasses and other vegetation. Bacterial action in the rumen enables grazing animals to access nutrients locked inside plant cells.

A valuable by-product of bacterial activity in the rumen is that buffalo are provided with their own internal heating system. They digest such large volumes of plant matter that the rumen bacteria generate enough heat to maintain a constant body temperature of about 104° F (40° C). This has doubtless contributed to the ability of bison to live in the severe continental climates prevailing in the plains of North America and in the cold central European forests. The digestive processes of the rumen bacteria are aided by prolonged chewing of the semidigested and regurgitated plant material known as cud. To assist with chewing the cud, all members of the tribe Bovini have specialized ridged cheek teeth which aid in breaking down tough plant material.

Safety in numbers

Wild and domesticated buffalo are in general highly sociable. They form herds ranging in size from small, tightly knit groups of 10 to 20 related individuals to huge gatherings of hundreds and even thousands of animals, which come together during migrations or simply to feed and breed. This natural herding instinct affords protection from predators. Older, more experienced members of the herd act as lookouts,

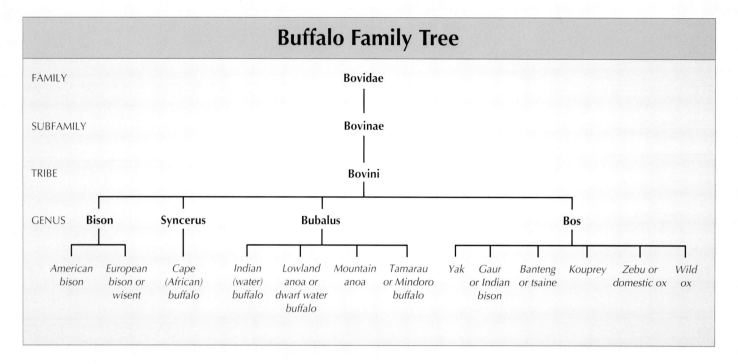

Buffalo Family Tree

FAMILY			Bovidae				
SUBFAMILY			Bovinae				
TRIBE			Bovini				

| GENUS | **Bison** | | **Syncerus** | **Bubalus** | | | | **Bos** |

| American bison | European bison or wisent | Cape (African) buffalo | Indian (water) buffalo | Lowland anoa or dwarf water buffalo | Mountain anoa | Tamarau or Mindoro buffalo | Yak | Gaur or Indian bison | Banteng or tsaine | Kouprey | Zebu or domestic ox | Wild ox |

giving a warning call when they sense danger. The usual response is to flee the scene as a group, with any young and weak animals safely in the midst of the herd. Although buffalo are not fast-moving, by sticking together they present a confusing target to potential attackers.

Not all buffalo have a communal lifestyle, however. Some old male Cape buffalo are loners and rely on their size to deter predators. The three relatively small Asian buffalo—the mountain and lowland anoas and the tamarau—often lead solitary lives because their enclosed forest habitat is not ideal for movement in herds. Even these species may form family units comprising a mother, calf and older offspring.

Communication

Buffalo indicate their social status and breeding condition to other members of the herd in a variety of ways. In particular the animals use specific postures and snorting calls to convey vital information about their current status. Buffalo also have an acute sense of smell, and subtle differences in the chemical composition of their urine and body odors are central to communication between herd members.

When male buffalo fight one another for the right to mate, permanent physical factors such as the animals' impressive dimensions and the size and shape of their horns are important. The Cape buffalo's horns may cover its entire head to form a bony helmet that provides a measure of protection when males clash their heads in mating disputes. Large horns are also sported by male American and European bison, Indian buffalo, banteng, yak and gaur. The tamarau and the two anoas have comparatively slender, pointed horns suited to life in dense vegetation, and look like miniature Indian buffalo.

Banteng have a glossy, rich brown coat and distinctive white "stockings." They roam much of Southeast Asia, but are threatened by deforestation and by hybridization with domestic cattle.

Buffalo are not territorial but occupy home ranges that meet all their needs. Cape buffalo (above) require ample grazing, shade from the midday sun and access to a muddy wallow.

decline in numbers was the result of decades of hunting by humans. The European bison has declined to an even greater extent; it used to wander in the extensive forests of Europe and western Asia in large herds, but is now an endangered species.

The Indian buffalo, in contrast to its relatives, has become a beast of burden and is an essential part of the agricultural systems of many developing nations. It pulls plows and carts and carries heavy loads. As a result of its importance to farmers, this species is now one of the most abundant large mammals of all, numbering well over 100 million individuals in captivity and in the wild. Indian buffalo have even been introduced to Australia, where they have escaped from captivity; about 250,000 animals now live in a feral state in remote parts of the country.

Most of the other buffalo are today under threat from overhunting, habitat loss and competition with domestic cattle. In Africa the Cape buffalo has suffered population declines caused by hunting and by culls to prevent the buffalo from grazing the same

Horns are not solely connected with male competition, because the females of all buffalo species also possess them. Female horns tend to be smaller and less developed than those of males, but nevertheless are formidable weapons against predators. Buffalo horns, regardless of sex, effectively increase the size of the skull, although they are relatively lightweight. Unlike the antlers of deer, they are not shed annually.

Close association with humans

The first buffalo appeared about 20 million years ago, probably in what are now southern Europe and Mongolia. Vast herds are thought to have been commonplace. Buffalo have been heavily exploited by humans since prehistoric times for their meat, bones and hides, and more recently, as domesticated animals. The changing fortunes of all the modern species of buffalo have become inextricably linked with humans.

Once among the most numerous large mammals in North America and Europe, the two species of bison are now found only in severely restricted parts of their former ranges. North American bison have been reduced to remnant populations in a few national parks and private collections. The catastrophic

land as farmers' cattle. Cape buffalo populations have now stabilized within a much restricted range in the uncultivated areas of the continent. Less fortunate are the anoas, kouprey, banteng and tamarau, all of which are now endangered. The kouprey, a relative of domestic cattle that was not discovered by science until 1937, faces possible extinction. Sometimes known as the Cambodian forest ox, it is native to Cambodia, Laos and Vietnam, and the recent history of war in this region has had a devastating impact on the species.

Another contributing factor to the steep decline in buffalo numbers is the fact that buffalo carry a wide variety of diseases such as rinderpest, brucellosis and tuberculosis that can be transmitted to domestic cattle. Even when domestic stock are vaccinated, wild buffalo populations can act as reservoirs for the diseases. For this reason disease control strategies often involve culling wild buffalo as well as implementing vaccination programs for domestic animals.

For particular species see:
- BISON • CAPE BUFFALO • GAUR
- INDIAN BUFFALO • YAK • ZEBU

BULBUL

The common bulbul is widespread and typical of its family. All bulbuls are small in size, with short wings and long tails.

BULBULS ARE A FAMILY of 20 genera, related to the babblers and native to the warmer, tropical and subtropical regions of the Old World. The appearance of some bulbul species varies across their geographical range; others are difficult to tell apart. Consequently, classification is not straightforward, and some authorities recognize a different number species from others. The name bulbul is of Arab origin, and is also Hindi and Persian for a particular species, the white-cheeked bulbul, *Pycnonotus leucogenys.*

Bulbuls are small birds, ranging from the size of a house sparrow to that of a thrush. The wings are short and the tail is comparatively long. Many species have a rather weak flight and so do not generally migrate. An exception is the migratory brown-eared bulbul, *Hypsipetes amaurotis,* of Japan. Around the base of the bulbul's bill are well-developed bristles, which may help guide food into the mouth. Bulbuls also have a patch of hairlike feathers on the back of the neck. Some species have a distinct crest.

The plumage of the sexes is much the same, and is rarely very brilliant except in species with patches of yellow, red or white. African members of the genus *Phyllastrephus* are often called brownbuls or greenbuls, and have greenish or brownish backs and yellowish underparts. In these species the females are smaller than the males and have shorter bills.

Gregarious songbirds

Outside the breeding season, bulbuls are gregarious, living in noisy flocks, and sometimes mixing with other species. As well as producing chattering noises, bulbuls also create distinctive whistlelike songs. The song of the white-cheeked bulbul is strongly reminiscent of the song of the European nightingale. Because many species are songsters and are comparatively easy to tame, bulbuls have become popular cage birds, especially in Asia.

Most bulbul species are forest-dwellers, but others live in sparse woodland and visit orchards. Over various parts of the bulbuls' range certain species have established themselves as birds of gardens and towns, the equivalent of blackbirds and house sparrows in Europe. Apart from the greenbuls, which are mainly insect-eaters and have stouter bills, the bulbuls feed on fruits and berries, supplemented with insects and spiders. Some species drink flower nectar. Bulbuls usually forage in the upper parts of trees, but will occasionally search for insects on the ground. Insectivorous species sometimes also catch insects in flight, in the style of flycatchers. In some areas bulbuls will descend in numbers on fruit and crops and strip trees bare. In the mountains of India, the black bulbul, *H. madagascariensis,* is known as a pest in cherry orchards.

Elaborate courting rituals

The male bulbul courts the female by puffing out his body feathers to display the colored patches that are often at the base of the tail. The female responds by chirping and quivering her wings. If she has a crest, she will retract it.

The bulbul's nest is cup-shaped and is built in a tree fork, usually near the ground. It is made of grass, pine needles, bamboo and other readily available materials, and is lined with grass or fungi. The nest's loose construction enables tropical rains to drain through it easily. The hen lays two or three eggs, rarely up to five, at one-day intervals. Incubation then proceeds for 10–15 days, and is usually carried out by the female, which is fed by the male. Both parents feed the chicks. Mongooses, crows, magpies, owls and hawks often take the broods of bulbuls. The adult bulbuls will, however, combine to defend

COMMON BULBUL

CLASS	**Aves**
ORDER	**Passeriformes**
FAMILY	**Pycnonotidae**
GENUS AND SPECIES	***Pycnonotus barbatus***

ALTERNATIVE NAME
African bulbul

WEIGHT
Up to 1¾ oz. (50 g)

LENGTH
**Head to tail: 7½ in. (19 cm);
wingspan: 10–12 in. (26–30 cm)**

DISTINCTIVE FEATURES
**High crown; long tail feathers; short wings;
hairlike bristles on nape of neck; dusky
colored with bright undertail**

DIET
**Mainly fruits and berries, with some shoots
and buds; also nectar, insects and spiders**

BREEDING
**Age at first breeding: 1 year; breeding
season: mid-May–August; number of eggs:
2 or 3; incubation period: 12–14 days;
fledging period: 12–14 days; breeding
interval: usually 2 or more broods per year**

LIFE SPAN
Not known

HABITAT
**Forests, thickets and shrublands; other
fertile places, such as gardens, orchards and
wooded streams**

DISTRIBUTION
**Africa, except for Sahara Desert and parts of
north, and from Middle East through India
to Southeast Asia; southeastern Australia**

STATUS
Generally common

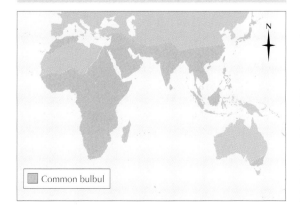

Common bulbul

the young against these predators. It is also common for the offspring from previous broods to help their parents feed and protect the current one, thus increasing the rate of breeding success.

Spreading suburbanites

Bulbuls have been carried to different parts of the world, and some have escaped from captivity, or have been deliberately released. Different species of bulbul are now found in the wild in Fiji, the Caribbean, New Zealand, Australia and the United States. Wild populations of the red-whiskered bulbul, *P. jocosus*, have formed in Sydney, Melbourne, Singapore, Mauritius and Florida; populations of the red-vented bulbul, *P. cafer*, have been found in the wild in Fiji and in Auckland, New Zealand. In Florida wild bulbuls were first seen in Kendall, just south of Miami. A few months later they had spread to Princeton, 35 miles (56 km) away, and in 3 years the population numbered 40 to 50 birds. Although in their natural range these birds are primarily forest-dwellers that became adapted to living in gardens, the immigrant bulbuls stayed in the suburbs, and did not venture out into the country. It is possible that only suburban gardens provided the birds with a suitable food supply.

When the red-vented bulbul first appeared in Auckland in 1952, future imports were made illegal to stop the species becoming a pest. The public was asked to report any bulbuls they saw so that the invading species could be controlled. However, the process of outlawing the import of this bulbul resulted in many captive birds being released on arrival at the ports, before customs officers could find them. The red-vented bulbul has since spread to southeastern Australia.

Originally from Asia, the red-whiskered bulbul has been successfully introduced to Fiji, Australia (in the 1880s) and Florida (in 1960).

BULLFINCH

When the chicks are very young the male bullfinch (left) brings seeds and insects to the female, which feeds the nestlings. Later both parents forage for the fast-growing brood.

THE BULLFINCH IS ONE OF the most distinctive of Europe's small birds; the chestnut red underparts of a perched male are unmistakable. However, bullfinches are secretive and the most common view is of the flashing white undertail of the bird in flight. Bullfinches are not often seen on the ground, where they move by hopping rather heavily. Usually they perch in the cover of trees, flying with a typically finchlike undulating flight from one tree to another. Throughout most of the year bullfinches live in pairs or small family parties. In the spring they may occasionally be seen in flocks, though rarely of more than 100 birds.

Changing status

Bullfinches are found across Europe and Asia, from Ireland and Britain east to Japan. In most regions they are restricted to coniferous or mixed woodlands, but in Japan and Britain they live in deciduous woodland and also visit cultivated ground.

The British subspecies, or race, of bullfinches is rather smaller and less brightly colored than the European race, which is only occasionally seen in Britain. Bullfinch populations increased in England from the 1940s to the 1970s, but since then have fallen again, perhaps due to the removal of mature hedgerows and to the introduction of more intensive agricultural practices. The decline may also be linked to a steep rise in the number of domestic cats, which prey heavily on many species of small songbird.

Nipped in the bud

Bullfinches feed on seeds and the buds of fruit trees, and are considered a major pest in orchards. Studies have shown that the amount of damage inflicted is related to the size of the species' natural food supply.

A bullfinch feeds on buds systematically, landing on the tip of a branch and slowly moving in toward the trunk, stripping the buds as it goes. When the bird reaches the older wood where there are fewer buds, it flies out to the tip of another branch. This is a highly efficient method of removing buds: the bullfinch can deal with up to 30 buds a minute. Orchards capable of yielding several tons of fruit can be stripped by bullfinches so thoroughly that only a few pounds can harvested.

It is only when supplies of seeds left over from the previous summer and fall are exhausted that bullfinches attack buds. Sometimes this may not occur until late spring, but when the last season's crop of seeds has been poor, buds may be taken throughout the winter.

Fussy eaters

In deciduous woodlands, bullfinches show a definite preference for the seeds of certain plants. Favorites include docks, nettles, privet, bramble, birch and ash; a very few seeds of other species are also taken. The seeds ripen in October, forming a food supply that has to last the bullfinch until the buds develop in spring. Birch and privet are preferred, but after all the others have

BULLFINCH

CLASS	**Aves**
ORDER	**Passeriformes**
FAMILY	**Fringillidae**
GENUS AND SPECIES	***Pyrrhula pyrrhula***

ALTERNATIVE NAMES
Common bullfinch; Eurasian bullfinch

WEIGHT
⅔–1⅓ oz. (16–38 g), varies according to subspecies

LENGTH
Head to tail: about 6 in. (15 cm); wingspan: 9–11½ in. (22–29 cm)

DISTINCTIVE FEATURES
Strong, broad-based bill; thickset neck; stout body; black cap; white rump. Male: reddish underparts, gray upperparts. Female: buffish underparts, grayish buff upperparts.

DIET
Adult: seeds of hardwoods and conifers; berries and buds. Nestling: insects and seeds.

BREEDING
Age at first breeding: 1 year; breeding season: late April–August; number of eggs: 4 or 5; incubation period: 12–14 days; fledging period: 12–18 days; breeding interval: usually 2 broods per year

LIFE SPAN
Up to 17 years

HABITAT
Woodland with thick undergrowth, orchards, hedgerows and gardens

DISTRIBUTION
Western Europe north to Scandinavia and east through Central Asia to Japan

STATUS
Common, but declining in some areas

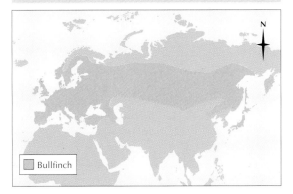

Bullfinch

The male, or cock, bullfinch can be identified by its chestnut red breast. Females are duller in color and have a grayish breast.

been eaten, ash forms the bulk of the bullfinch's food supply. The size of the crop of ash seeds varies enormously with, typically, a two-year cycle of abundance. One year the crop is plentiful, the next very few seeds are produced. It is in the lean years of shortages that bullfinches turn to buds to make up their diet and become a pest in orchards.

When bullfinches do turn to eating buds, some fruit varieties are targeted before others. The buds of dessert apples are eaten before those of cooking apples and striking contrasts may be seen in orchards where the two varieties of fruit are grown in alternate rows. The one will be stripped, the other a mass of blossom.

Parental care and predators

Nests of fine twigs with moss and lichen and a lining of intertwined roots are built in shrubs and hedgerows, usually only a few feet above the ground. Two clutches of four to five eggs are usual, and are laid from early May to mid-July. However, if food is plentiful, breeding may continue until October. The cock bullfinch helps the hen to incubate the eggs and also feeds her while she is on the nest. Chicks hatch out in 2 weeks and are fed on a mixture of seeds and insects, with the latter predominating at first. The cock provides all the food when the young are newly born, but later both parents collect food for the chicks, which fledge in just over 2 weeks.

One study of bullfinches found that only one-third of all clutches in woodland and two-thirds of those in farmland were successful. The rest were eaten by predators, especially jays, which are more numerous in woodland, thus accounting for the greater survival of young bullfinches in farmland. Magpies, ermines and weasels also prey on bullfinch broods.

BULLFROG

The bullfrog favors marshes, ponds and slow-flowing rivers. It rarely moves far from water except during unusually wet weather.

THE BULLFROG IS ONE of the largest species of North American true frogs and may grow up to 8 inches (20 cm) long. The tadpoles are the largest in North America, commonly reaching a size of 1½–3½ inches (4–9 cm). The bullfrog is named for its deep, resonant croak. It is native to the United States, and is found east of the Rockies, and on the northern borders of Mexico. The species has also been introduced to the western states, as well as to Canada, Hawaii, Mexico, Cuba, Jamaica, Italy and Japan. The bullfrog has flourished in Europe, often at the expense of the native wildlife.

The North American bullfrog is similar in appearance to the much smaller edible frog of Europe; its skin is usually smooth but sometimes it is covered with small tubercles. On its upperparts the frog is usually greenish to black, sometimes with dark spots, while the underparts are whitish with tinges of yellow. The females are more highly spotted than the males and are browner in color; males have a yellower throat than females. It is also possible to distinguish the sexes by comparing the size of the eye and the eardrum, which is exposed. In females they are equal; in males the eardrum is larger than the eye.

Bullfrogs are usually found in or near water, but venture further afield during very wet weather. They prefer to live near ponds, marshes and slow-flowing streams, and may be found lying along the water's edge under the shade of shrubs and reeds. In winter they hibernate near water, under logs and stones or in holes on the banks. The duration of their hibernation is dependent on the climate. Bullfrogs are often the first amphibians in an area to retire at the approach of winter and the last to emerge in the spring. In the northern parts of their range bullfrogs usually emerge around the middle of May, but in Texas, for example, they may emerge in February in mild weather. In the southern areas of their range they may not hibernate at all.

Voracious appetite

The bullfrog feeds mainly on insects, earthworms, spiders, crayfish and snails. Many kinds of insects are caught, including grasshoppers, beetles, flies, wasps and bees. The bullfrog captures its prey by lying in wait and then leaping forward as its victim passes. Its tongue is rapidly extended by muscular contraction and wraps around the prey. The frog then submerges to swallow its victim. A voracious carnivore, the bullfrog takes active adult insects as well as the slow-moving larvae and immobile pupae.

The bullfrog also takes larger prey, including other frogs, fish, small terrapins, newly hatched alligators, turtles and small mammals, such as mice and shrews. Small garter snakes and even venomous coral snakes are eaten. There is one recorded case of a coral snake measuring 17 inches (43 cm) being taken by a bullfrog. The bullfrog also captures small birds, especially ducklings. Even swallows, flying low over the water, are not safe from the bullfrog's voracious appetite and considerable leaping

BULLFROG

CLASS	**Amphibia**
ORDER	**Anura**
FAMILY	**Ranidae**
GENUS AND SPECIES	***Rana catesbeiana***

ALTERNATIVE NAMES
American bullfrog; jug o' rum (archaic)

WEIGHT
2–9 oz. (50–250 g)

LENGTH
3½–8 in. (9–20 cm)

DISTINCTIVE FEATURES
Large size; usually greenish olive, paler below; female browner than male

DIET
Mainly insects, earthworms, snails, spiders and aquatic animals, including tadpoles and crayfish; larger prey includes turtles, snakes, small mammals, birds and other frogs

BREEDING
Age at first breeding: 2–3 years (male), 3 years (female); breeding season: February–August, earliest in south; number of eggs: up to 25,000; hatching period: about 7 days; larval period: up to 2 years; breeding interval: 1 or 2 clutches per year

LIFE SPAN
Up to 16 years in captivity

HABITAT
Fresh water with dense aquatic vegetation, often in prairies, chaparral, farmland and woodland

DISTRIBUTION
U.S. east of Rocky Mountains. Introduced to western U.S. and Canada, Hawaii, Mexico, Cuba, Jamaica, Italy and Japan.

STATUS
Common

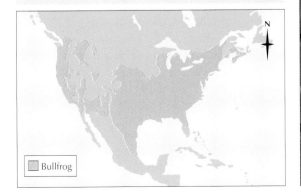

Bullfrog

ability; adult bullfrogs can leap over 3 feet (0.9 m). This indicates that the bullfrog is both an opportunistic feeder and a powerful predator.

Unusual mating call

When the water temperature reaches about 70° F (21° C), mating takes place. This condition arises about February in the south of the bullfrog's range, and in June or July in the northern parts. At night the males move out from the banks to call, while the females stay inshore. They join the males only when their eggs are ripe.

The bullfrog utters a distinctive hollow, booming call during the mating season. The sound is made three or four times within a few seconds and repeated after an interval of about 5 minutes. The call is produced by air being passed back and forth along the bullfrog's windpipe, from lungs to mouth, with the nostrils closed. Some air enters the air sacs in the floor of the mouth which swell out like balloons and act as resonators, amplifying the sound so that the noise can be heard ½ mile (0.8 km) away. The bullfrog's call sounds similar to the words "jug o' rum" enunciated in a gruff tone, and gave rise to one of the bullfrog's traditional names.

After mating the female bullfrog lays 10,000 to 25,000 eggs, which float in a film on the surface of the water among water plants. The size of the clutch depends on the size of the female. Including its envelope of jelly, each egg is just over ½ inch (1.5 cm) in diameter. It is black above and white below. The eggs usually hatch within a week of being laid. If the temperature is low, however, they may take over 2 years to metamorphose into an adult frog, by which time they are 2–3 inches (5–8 cm) long. The young bullfrogs feed on algae and decaying vegetation, though they occasionally take small pond animals. Approximately 2 years later the young are almost fully grown and are ready to breed.

It is possible to distinguish male and female bullfrogs by their coloring. The females are browner and more highly spotted than the males.

BULLHEAD

The male bullhead guards the clutch of eggs until they hatch, aerating them by making fanning movements with his tail.

SEVERAL DIFFERENT KINDS OF fish are called bullheads, but the species considered here is the freshwater fish, *Cottus gobio*, that lives in European rivers. At one time this fish was known as miller's-thumb, because of its resemblance to the broad, splayed shape of the end of a miller's thumb, caused by constantly rubbing flour between thumb and fingers to gauge its texture. More recently, however, the species has been given the name of bullhead.

The European bullhead has a broad, flattened head, rounded in front, and a stout body that tapers to the tail. Its dorsal fins are large and spiny and there is a stout spine on the gill cover. The bullhead is variable in color, usually greenish yellow to brown with dark bars and blotches, and it can change color to harmonize with its background when resting on a stony bed.

Most members of the family Cottidae, to which the European bullhead belongs, are marine fish. However, a number of North American cottids are, like the bullhead, found in freshwater. These North American species are known as sculpins. Among the unrelated fish called bullheads are the North American catfish of the family Ictaluridae and a small species of goby found in New Zealand.

Unobtrusive fish

The bullhead lives in the shallows in clear, gently flowing streams and in lakes, hovering over the gravel or beneath stones, and moving little except to feed or when disturbed. If surprised, it is capable of darting away at speed. The function of the bullhead's broad fins is to give it this high degree of maneuverability among the stones on the bed of the stream or lake. However, the fish is not capable of swimming any great distance.

Bullheads are nocturnal fish and spend much of the day hiding in cracks and crevices. A fish of either sex will drive off a smaller fish that

BUMBLEBEES

PHYLUM	**Arthropoda**
CLASS	**Insecta**
ORDER	**Hymenoptera**
SUBORDER	**Apocrita**
FAMILY	**Apidae**
GENUS	***Bombus***
SPECIES	**More than 200**

ALTERNATIVE NAMES
Humblebee; carderbee (some species only)

LENGTH
⅓–1 in. (0.7–2.5 cm); queen largest, worker smallest

DISTINCTIVE FEATURES
Larger and more hairy than honeybee; usually black with variable amount of yellow or orange banding; some species have brown or reddish tails

DIET
Flower nectar and pollen

BREEDING
Breeding season: spring, summer and fall; number of eggs: 200 to 400 per queen

LIFE SPAN
Adult: 2–12 months, according to species

HABITAT
Grassland and woodland, near flowers. Nest sites include areas of long grass, moss, dead timber, mouseholes and inside buildings.

DISTRIBUTION
Mainly temperate zones of Northern Hemisphere, also parts of South America and Southeast Asia. Introduced to Australia and New Zealand.

STATUS
Relatively common

Bumblebee workers are infertile females with only two roles in life: to gather food for the colony and to tend its grubs. These workers are white-tailed bumblebees, Bombus lucorum.

The life of a bumblebee colony begins in the fall. A young female bumblebee leaves the nest, mates and finds a sheltered spot where she can spend the winter hibernating. When spring comes—the actual date depends on both the species and weather conditions—the young queen emerges and suns herself until she is fully active. She gathers pollen and nectar from spring flowers. Pollen is very rich in protein, which the queen needs to form eggs to be laid later.

Soon the queen starts looking for a suitable place in which to build her own nest. She might take over the abandoned nest of a wood mouse, vole or hedgehog, or she might pick a disused bird's nest, a bale of hay or even a discarded mattress. Favorite sites include along hedgerows and banks and in old, neglected corners of fields and gardens. Today it is more difficult for queen bumblebees to find this kind of nest site because modern intensive and mechanized farming demands that these unproductive corners be plowed. The result is that bumblebees, which are involved in the pollination of many crops, are becoming scarcer.

The queen usually builds her nest at the end of a tunnel. The tunnel may be several feet long if she has used an old mouse nest, but some species prefer tunnels only a few inches long and so build their nests in thatch or similar places. A few species of bumblebees, known as carderbees, even manage to build their nests on the surface of the ground by combing grass and other material into a tight, closely woven ball. If the queen bumblebee has taken over the nest from a former occupant, there is always plenty of nest material readily available. If not, suitable material must be

and help look after her next brood. Certain wasps, honeybees and ants are well-known social insects. Thousands of them may live in one nest. Most of them are sexless or, rather, under-developed females. Known as workers, they care for the breeding female, or queen, and for her eggs and larvae. Bumblebees are social insects, too, but their communal life is not as well developed as in other social species. Their colonies have fewer workers and all but the queens die in the fall or early winter.

A male and female bumblebee, Bombus terrestris, mating. Male bumblebees die soon after mating, which takes place in the fall. Fertilized females hibernate and establish a new colony in the spring.

Before long the first brood emerges from the cocoons as fully developed workers. They spend up to two days drying out while their wings expand and harden. The workers are then ready for the duties of venturing out to collect food and tending the next batch of larvae. An entire colony of several hundred workers is soon built up. The queen bumblebee, however, is never reduced to a helpless egg-laying machine, as happens with ants and termites. She can still make egg cells and feed larvae, although she hardly ever leaves the nest to forage.

Toward the end of summer, some of the eggs produce males and fertile females instead of the nonreproducing female workers produced thus far in the season. The males develop from unfertilized eggs. The females—the next generation of queens—appear at first to be exactly the same as the sterile female workers but grow much larger and finally leave the nest to mate. Male bumblebees use their unusually large antennae to locate these fertile females. Once the old queen has produced the males and the new queens, she stops laying worker eggs and the whole colony gradually dies out. After mating the male bumblebees also perish. When winter arrives only the young queens are left to survive through to spring.

Many predators

Bumblebees have a host of predators, large and small. Among the most important are shrews, mice and insect-eating birds, such as bee-eaters. Badgers, skunks and other mammals will dig up nests both for the honey and for the bees themselves. Skunks, for example, have been observed scratching at nests until the inhabitants fly out and are caught in the forepaws. Robber flies grapple bumblebees with their legs and suck the bees' blood, while certain species of mite live inside the air sacs, or "lungs," of bumblebees and feed on their juices.

Among the insect parasites of bumblebees are wax moths and cuckoo bees. The wax moth lays its eggs inside bumblebee nests and its caterpillars ruin the egg cells by burrowing through them. Cuckoo bees are close relatives of bumblebees, but because they do not have pollen baskets they are unable to provision nests of their own. Instead they invade the nests of bumblebees and lay their eggs there. Cuckoo bee eggs develop into males and females, but not workers, and the adults depend on the bumblebee workers.

When attacked a bumblebee attempts to defend itself by biting and stinging. It rolls onto its back with its jaws open and sting protruding and sometimes squirts venom into the air. Its sting is not barbed like the sting of a honeybee, so it can be withdrawn from the corpse of an attacker and used again.

gathered from elsewhere. The queen fashions this into a small inner chamber lined with fine grasses and roots. She often stays there for a day or two, drying it out with her body heat. Insects are cold-blooded, but the larger ones generate enough heat, especially with their flight muscles, to keep their bodies a few degrees above air temperature outside the nest.

By now the eggs are developing inside the queen bumblebee. In most species the queen proceeds to make one or more egg cells out of wax secreted from between the plates on the underside of her abdomen. She collects pollen and stores it in the cell or cells, in each of which a variable number of eggs are laid. Each cell is then covered with a cap of wax. The queen spends some of her time settled on the eggs to keep them warm. She occasionally leaves to feed, bringing home any surplus nectar, which she stores in a "honey pot" near the nest entrance. This is also made of wax and is ¾ inch (2 cm) high and ½ inch (1 cm) across. It stores food for when the weather is too bad for the queen to leave and forage.

When the larvae hatch out of the eggs, they are helpless grubs with little in the way of legs or sense organs. They do nothing except feed on the pollen that has been stored in the egg cell and on the mixture of nectar and pollen that the queen regurgitates to them through a hole in the top of the cell. The larvae grow remarkably quickly on this diet and eventually shed their skins several times, spin a cocoon and pupate.

At this point the queen carefully removes the wax from around the cocoons and remolds it into new egg cells. She rests these on top of the cocoons and lays the next batch of eggs in them.

BUNTING

IN THE OLD WORLD the name bunting is used for members of the subfamily Emberizinae, whereas the North American representatives of this subfamily are usually known as sparrows and finches. In the United States the term bunting is reserved for the colorful members of another, closely related subfamily, Cardinalinae, which includes the cardinals. The Old World buntings are finchlike birds with short, conical bills used for crushing seeds, and have a slighter build and longer tail than American "buntings."

One of the most widespread buntings is the snow bunting, *Plectrophenax nivalis*, which has a circumpolar breeding distribution. It is among the most northerly breeding of all passerines (species of the order Passeriformes). Snow buntings retreat south in advance of the bitter Arctic winter; in North America they see out the coldest months on the prairies of southern Canada and the northern United States and in coastal regions as far south as northern California and Virginia. Elsewhere snow buntings winter in northern and central Europe and across central Asia.

The male snow bunting has a piebald (black-and-white) breeding plumage. Its close relative, McKay's bunting, *P. hyperboreus*, is almost all white, with black restricted to the wing tips. Unlike the snow bunting, this species has a very restricted range, comprising the western coasts of Alaska and a few islands in the Bering Sea. The Lapland bunting, *Calcarius lapponicus*, or longspur, breeds on the Arctic tundra, migrating to the prairies of the midwest for the winter.

Common but often overlooked

Buntings are widely distributed over Europe, Asia and Africa, most of them preferring open country, such as tundra, moors, fields, bush and grasslands. A few species live in light woodland or on the fringes of forests, others in reed beds, but buntings are not often found in dense forests. Many buntings are migratory, moving south in the fall and returning to their breeding grounds in spring. They migrate at night to avoid diurnal (day-active) predators and so that they can replenish depleted energy reserves by intensive feeding during the daylight hours.

Outside the breeding season many buntings form large flocks, often joining with finches and other species of bunting, for both feeding and roosting. A group of buntings is more likely than a single bird to spot the approach of a predator, such as a hawk, and when the mass of birds takes off at once it presents a more difficult target. Moreover, buntings often have relatively

dull plumages in shades of brown, gray and black. This inconspicuousness is heightened by wariness. For instance, the ortolan bunting, *Emberiza hortulana*, of northern and eastern Europe and the eastern Mediterranean is a dull little bird with secretive habits. On the other hand, the house bunting, *E. striolata*, shows little fear of humans, and even nests in houses.

Buntings feed on the seeds of plants, especially grasses, and on invertebrates, including flying insects, snails and crustaceans. The proportion of the different foods varies according to species and season. The ortolan bunting also feeds on locusts while migrating across Africa.

Male rivalry

The arrival of snow buntings at their breeding grounds is traditionally a time of excitement for Inuit people, as it heralds the coming of spring. The male birds arrive first, and spend their time feeding and roosting. Later they start disputing the boundaries of nesting territories. The males sing from lookout points on boulders and fly into the air while singing to inform rivals of their territory. If a strange male enters a territory, the

In fresh breeding plumage the male painted bunting, Passerina ciris, is among the most striking of all North American songbirds. The species is confined to the southern states.

The ortolan bunting is typical of the many well-camouflaged species in the bunting family.

owner chases it out, and when neighbors meet at the boundary, they indulge in a "pendulum flight." Each bird alternately advances and retreats so that the two birds fly backward and forward over the boundary as if linked by an invisible rod. Eventually the female snow buntings arrive and courtship begins. A notable feature of bunting courtship is a chase, in which the male rushes after the female in a headlong, twisting and turning pursuit until he catches her, whereupon a fluttering and tumbling brawl develops, both birds falling to the ground.

Bunting nests are usually cup-shaped, or domed in tropical species, and are made out of grasses, fine roots, moss and lichens. Most species nest a few feet off the ground. The Old World reed bunting, *E. schoeniclus*, often builds its nest in reeds and sedges over water, while others build their nests in tree holes or on the ground among rocks and hummocks. The eggs number two to six, and are incubated for 12–13 days. In some species, both sexes share incubation and feeding of the young; in others, these duties are left to the female alone.

Polygamy has been recorded in an exclusively Old World species of bunting, the corn bunting, *Miliaria calandra*. In Cornwall, southwestern England, 15 male corn buntings were observed to mate with a total of 51 females between them, some mating with as many as 7 females. The males did not help to rear the families, but spent time on the lookout for intruders. The females built nests and raised chicks, often within a few yards of one another.

The overriding aim of an individual animal is to maximize its reproductive potential. Generally, female songbirds achieve this by favoring males that have good genes and provide parental care; males, on the other hand, sometimes trade the provision of parental care for new mating opportunities. The outcome of this conflict between the sexes depends on many factors, including ecological ones such as food availability.

SNOW BUNTING

CLASS	**Aves**
ORDER	**Passeriformes**
FAMILY	**Emberizidae**
SUBFAMILY	**Emberizinae**
GENUS AND SPECIES	***Plectrophenax nivalis***

WEIGHT
1–1¾ oz. (28–50 g)

LENGTH
Head to tail: 6–6½ in. (15–17 cm); wingspan: 12½–15 in. (32–38 cm)

DISTINCTIVE FEATURES
Stubby bill. Male, summer: snow white plumage with black back and wing tips; winter: upperparts, nape and crown become sandy brown. Female: duller, with brownish gray crown and upperparts.

DIET
Mainly seeds; also insects in spring and summer

BREEDING
Age at first breeding: 1 year; breeding season: mid-May–July; number of eggs: 4 to 6; incubation period: 10–15 days; fledging period: 10–14 days; breeding interval: 1 or 2 broods per year

LIFE SPAN
Up to 4 years

HABITAT
Summer: tundra and rocky scree; winter: rough grassland and coasts, especially salt marshes and sand dunes

DISTRIBUTION
Breeds in circumpolar range in far north; winters in mid-latitudes of Northern Hemisphere

STATUS
Fairly common

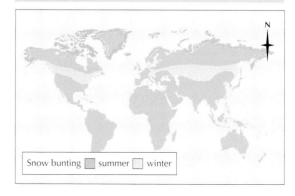

Snow bunting ■ summer □ winter

BURNET MOTH

THE BRIGHT COLORS of burnet moths make them extremely conspicuous. These insects are small to medium-sized, with thick bodies, and their hind wings are shorter than their forewings. The antennae are often described as clubbed because they are thickened near the tip. Most species of burnet moth are red and black, and freshly emerged individuals have a deep metallic blue or green sheen in the black areas.

There are many different color variations among the species of burnet moth, often featuring yellow or white patches and spots. In the striking *Zygaena carniolica* of central and southern Europe and western Asia the red spots are surrounded by pale rings. The eastern grapeleaf skeletonizer, *Harrisina americana*, is characterized by slender black wings, a long, black body and a red or orange collar behind the head. This species is widespread in the eastern United States. Its name is derived from the fact that its larvae feed off the leaves of a plant until only a skeleton is left.

Burnet moths belong to the family Zygaenidae, which includes many species with metallic-looking colors and thin, long hind wings. One subfamily of the Zygaenidae contains a southern Asian species, Doherty's longtail, *Himantopterus dohertyi*, in which the hind wings form long, trailing streamers behind the forewings. Most of the brightly colored Zygaenid moths, including all of the burnet moths, are day fliers. The combination of bright coloration, diurnal (day-active) behavior and clubbed antennae means that burnet moths are often mistaken for butterflies.

Colonial breeders

Burnet moths occur in Europe, temperate Asia and Africa, mainly in lowland areas. Some of the species found around the coasts of the Mediterranean are often present in immense numbers. Burnet moths usually live in colonies, frequently occupying only part of a hillside or a single meadow. In a flourishing colony, the moths are abundant, sometimes to the extent that half a dozen or more are seen on each flower head. They are most active when the sun is shining, although their flight is slow.

The adult moths feed on the nectar of flowers, sitting on flower heads and probing the nectaries with their long, tubular tongues. The larvae, or caterpillars, feed on the leaves of various food plants. These vary between different species but generally include thistles, vetches, trefoils, clovers and thyme. Burnet moth caterpillars usually feed during the late summer, hibernating through the winter and completing their growth during the spring and early summer of the following year. An exclusively Scottish species, the mountain burnet, *Z. exulans*, is exceptional in taking 1–4 years to complete its life cycle. Only a short time is spent as a pupa

The markings on burnet moths, such as this six spot burnet, Zygaena filipendulae, warn predators of the moths' unpleasant taste.

and the adult moths probably live only 2 or 3 weeks. The caterpillars of burnet moths are thick-bodied and grublike, and greenish or yellowish in color with a regular pattern of black spots. They also bear small protuberances with tufts of fine, short hairs arising from each bump.

The pupa is enclosed in a characteristic spindle-shaped cocoon of parchmentlike silk and is often yellow or white in color. The cocoons are usually attached to the stem of a plant, those of the more common species being conspicuous and easy to find. After the adult moth has hatched, the empty black pupa shell sticks out of the cocoon.

Noxious taste

The habits of burnet moths would seem to make them an easy prey for birds and other insect-eating animals. The moths are slow-flying and conspicuous; they make no attempt to hide, and little attempt to evade capture. However, most burnet moths have chemical defenses that render them distasteful or even toxic. It is the resultant lack of natural predators that explains why these moths are so easy to approach and catch.

Since burnet moths are ill-tasting and to some degree poisonous, a bird that has pecked one of them is not likely to attack another. It is to the moths' advantage to be brightly colored and easily recognizable so that predators have no difficulty in learning to avoid them. Their poison is discharged from the neck, mouth and feet in

Chemical protection enables burnet moths to feed out in the open without risking capture.

the form of a fluid and contains histamine and prussic acid, or hydrogen cyanide. In one experiment a captive-bred rook, *Corvus frugilegus*, a member of the crow family, was offered a burnet moth to see whether it would be accepted or rejected. The rook picked up the moth and dismembered it by severing the wings and biting off the head. Within seconds the bird was running in an agitated manner, stopping every so often to bite at cool grass blades and to rub its bill on the grass and bare earth. From time to time the rook spread its wings in the manner that is associated with birds that have something acrid in their mouth.

BURNET MOTHS

PHYLUM	**Arthropoda**
CLASS	**Insecta**
ORDER	**Lepidoptera**
FAMILY	**Zygaenidae**

GENUS AND SPECIES **Many, including variable burnet moth, *Zygaena ephialtes*; Provence burnet moth, *Z. occitanica*; and fire grid burnet moth, *Arniocera erythropyga***

LENGTH
Wingspan: variable and Provence burnet moths, 1¼–1½ in. (3–4 cm); fire grid burnet moth, 1–1¼ in. (2.5–3 cm)

DISTINCTIVE FEATURES
Antennae thickened near tip; most species red and black with dark green or bluish sheen to black areas; some species have pale yellow or white patches and spots

DIET
Adult: flower nectar. Larva: food plant depends on species; variable burnet moth: crown vetch, *Coronilla varia*; Provence burnet moth: plants of the pea family.

BREEDING
Breeding season: summer

LIFE SPAN
Adult: usually 2–3 weeks

HABITAT
Mainly grasslands, heaths and marshes in lowland areas

DISTRIBUTION
Eurasia and Africa

STATUS
Generally common; most plentiful in Europe

BURROWING OWL

THE BURROWING OWL stands 9 inches (23 cm) high, about the same size as the little owl, *Athene noctua*, of Europe and Asia. The upperparts are brown with white markings, while the underparts are off-white and bear darker marks. It is largely a terrestrial creature, perching and hunting on the ground, and accordingly has a very short tail and proportionately long legs compared to the larger forest owls, such as the screech owl, *Otus asio*, of North America. The burrowing owl's habit of bobbing up and down is characteristic of the species. As its name implies, it nests in holes in the ground rather than in trees. These holes are sometimes excavated by the owls themselves but are often adapted from dens abandoned by burrowing mammals.

Threatened by agriculture

At one time the burrowing owl was common on the plains and prairies of North America, but the steady encroachment of agriculture has greatly restricted the range of the species. Plowing ruined the owls' burrows, and many owls were shot because of the chance that horses and cattle might break their legs in the holes. Families of burrowing owls have also been killed indiscriminately by the poison gas used against ground squirrels. However, providing that the terrain is not altered too greatly, the burrowing owl is under no real threat. Farmers have realized that, like other species of owl, it is mainly beneficial to humans because it preys on pests such as mice and insects.

The burrowing owl ranges from the Pacific coast of North America east to Minnesota and Louisiana. It breeds as far north as British Columbia and Manitoba and south to Tierra del Fuego, though it does not occur in the forests of the Amazon basin. To the east it is found in Florida and on some of the Caribbean islands. In winter some burrowing owls migrate from the northern parts of the range and the Florida population disappears outside the breeding season.

Hunts by day

Like the little owl of Eurasia, the burrowing owl is more diurnal in its habits than most owls. Its vision is not believed to be as sharp as that of other owl species. The outcome of experiments on the keenness of sight in owls showed that whereas the barn owl, *Tyto alba*, could locate mice in light intensities equivalent to one candle located nearly 2,000 feet (610 m) away,

The burrowing owl's long legs and short tail are adaptations to its ground-dwelling lifestyle.

The burrowing owl's white and brown markings break up its outline, providing excellent camouflage.

BURROWING OWL

CLASS	**Aves**
ORDER	**Strigiformes**
FAMILY	**Strigidae**
GENUS AND SPECIES	***Athene cunicularia***

LENGTH
Head to tail: 9–10 in. (23–25 cm); wingspan: 19½–22½ in. (50–57 cm)

DISTINCTIVE FEATURES
Long legs; short tail; pale brown facial disc with prominent white eyebrows; sandy-brown above with white markings

DIET
Mainly small mammals including rodents, ground squirrels and young prairie dogs, and invertebrates such as large spiders, beetles and grasshoppers; large numbers of newly fledged birds in breeding season; also lizards, snakes and scorpions

BREEDING
Age at first breeding: 1 year; breeding season: most of year, depending on region; number of eggs: 7 to 9; incubation period: 28 days; fledging period: not known; breeding interval: 1 year

LIFE SPAN
Usually up to 9 years

HABITAT
Arid and semiarid open country including prairies, rough grassland and golf courses

DISTRIBUTION
Western North America and Florida south through parts of Central America and Caribbean to southern South America as far as Tierra del Fuego. Absent from much of Amazon Basin.

STATUS
Locally common; has declined in many agricultural areas

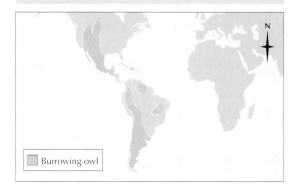

Burrowing owl

burrowing owls found it difficult to locate prey in far greater light intensities. Their sensitivity to light is in fact about the same as that of humans.

Hunting does take place at night, but is mainly carried out during the day. Burrowing owls pounce on prey, burying their talons into the victim's back. Prey species include mice, sage rats and other small rats, ground squirrels, young cottontail rabbits, chipmunks and even bats. Insects represent a large proportion of the diet, especially large beetles and grasshoppers, which are caught in the air in the talons. Large numbers of newly fledged birds, especially larks, are taken during the breeding season, but few other birds are eaten at other times. Other animals occasionally feature in the diet, and the remains of scorpions and centipedes are sometimes found near burrows. Crayfish are also eaten by burrowing owls living near water, which may simply take advantage of crayfish cast up on the banks or exposed by a drought.

Empty burrows taken over

Burrowing owls dig holes by scraping with their talons, but also frequently take advantage of burrows made by a wide variety of other animals. The holes of prairie dogs, groundhogs, American badgers, armadillos, foxes, skunks and many other burrowing creatures are taken over, and enlarged if necessary.

The nesting chamber, lined with grass, feathers, the stalks of weeds and other materials, is usually made at the end of a tunnel about

BURYING BEETLE

THERE ARE 15 SPECIES of burying, or sexton, beetle in North America, all of which are in the genus *Necrophorus*. Two of these, *N. vespilloides* and *N. investigator*, are also found in Europe. One of the largest species of burying beetle is about 1 inch (2.5 cm) long and black all over. The others are black with orange bands on their elytra (wing cases). Burying beetles are strong fliers and fly by night, often being attracted to lights. They feed on animal corpses and can detect the odor of rotting flesh up to ½ mile (0.8 km) away. The beetles bury small carcasses, such as the body of a small songbird, mouse or mole, and the females lay their eggs in them. If the body is too large for burial, the beetles feed on it and depart.

Burial chambers

Burying beetles have a unique and elaborate social system. Indeed, early observers thought that their complex behavioral patterns could be explained only if burying beetles were capable of reasoned thought. Several beetles may locate the same carcass, but usually only one male and one female will have the opportunity to breed there. Fights between beetles of the same sex determine which beetles have this privilege; normally it is won by the largest male and female. Similarly, if more than one species of burying beetle arrives at the carcass, generally only the larger species will breed. However, the smaller species, such as *N. vespilloides*, have adopted a strategy to thwart their larger relatives. Should several *N. vespilloides* beetles find a dead animal, they postpone fighting over it until they have cooperatively buried and concealed the carcass. In this way the smaller beetles reduce the chance of a larger species discovering the carcass.

The process of concealment begins with the beetles burrowing under the dead animal. They scrape earth away with the first pair of legs and throw it out to the sides with the two remaining pairs of legs. The beetles do this in such a way that the carcass, in sinking into the ground, folds on itself. As digging proceeds, earth thrown out by the beetles fills the space above. Eventually, at a depth of up to 8 inches (20 cm), a chamber slightly larger than the carcass has been excavated. This is known as the crypt. The beetles use their jaws to strip the carcass of its fur or feathers, which are used to reinforce the walls of the crypt.

When digging, the beetles may come across roots, which they bite through. If the carcass becomes entangled in grass and fails to fall into the hole, the beetles will cut through the grass.

This indicates recognition of a problem and the ability to search for a solution. Scientists have conducted experiments in which burying beetles are presented with a carcass which has been tied with string to prevent it from falling into a hole. The beetles bit through the string to free the carcass, overcoming the difficulty. Even wire does not defeat the beetles, for they respond by severing the limb to which the wire is attached.

Parental care

The female lays her eggs immediately after the carcass is buried, either within a tunnel or scattered around the carcass. The larvae hatch out in 2–5 days, during which time the parents roll the carcass into a perfect sphere, removing and eating any blowfly maggots. The beetles also add antibacterial secretions to the carcass to prevent it from decaying too quickly and to contain the powerful odor of decomposition, so that it is less likely to be detected by other beetles, ants and flies. The parents eat a bowl-shaped depression into the top of the ball of flesh where the larvae will congregate on hatching.

The adult burying beetle rubs the edge of its orange elytra (wing cases) against part of its abdomen to create a shrill noise. It uses this sound to guide its larvae to the cavity that it has constructed beneath an animal carcass.

Burying beetles feed on large animal corpses, then abandon them. Smaller carcasses, such as those of small rodents and birds, may be removed and buried for later consumption or for breeding.

BURYING BEETLES

PHYLUM	**Arthropoda**
CLASS	**Insecta**
ORDER	**Coleoptera**
FAMILY	**Silphidae**

GENUS AND SPECIES **About 15 North American species, including American burying beetle, *Necrophorus americanus***

ALTERNATIVE NAMES
Sexton beetle; gravedigger

LENGTH
Up to 1 in. (2.5 cm)

DISTINCTIVE FEATURES
Clubbed antennae; many species have orange bands on elytra (wing cases)

DIET
Carrion and (adult only) blowfly larvae

BREEDING
Age at first breeding: about 14 days after leaving pupae; hatching period: 2–5 days; larval period: about 15–18 days

LIFE SPAN
Adult: several months

HABITAT
Mainly woodland and forest; in Asia only in mountains

DISTRIBUTION
Mainly U. S., especially in north; Eurasia; absent from sub-Saharan Africa and Australia

STATUS
Generally common; *N. americanus*: critically endangered

When the larvae hatch they are guided to the crypt by the parents. In order to attract the larvae, the parents rub the edge of their elytra against a filelike structure on the abdomen to produce a shrill, singing noise. Each larva makes its way to the crypt either along the tunnel or by burrowing to it through the earth. Once in the crypt, the larvae sink their jaws into the pit at the top of the carrion. They feed directly from the flesh and, in most species, are able to complete development without further parental assistance. However, the parents tend to stay in the crypt, eating the carrion and regurgitating it in partly digested form for the larvae. The larvae compete with each other to take this liquid from the parents' jaws. The larvae of burying beetles are unusual in that they change form at each of the three molts preceding the pupal stage. To pupate, each larva bores through the earth to about 1 foot (30 cm) away from the crypt.

In almost all burying beetles, males as well as females stay with the young, at least for the early stages of development and both parents protect their larvae from other beetles. Highly developed parental care is rare in insects, which usually desert their young. It is unusual for male beetles to assist because they cannot be sure that the larvae are in fact their own; female beetles can store sperm from other males in their bodies until it is required to fertilize eggs. A female burying beetle can rear a brood alone if no male joins her at a carcass or if the male leaves. However, a single female is far less likely than a pair of beetles to successfully protect larvae.

Conservation in the United States

In common with the other members of the carrion beetle family Silphidae, the American burying beetle, *N. americanus*, performs the vital role of recycling decaying materials into the ecosystem. This beetle was once found through-out the United States east of the Rocky Mountains but has vanished from most of its former range. The species now occurs only in Oklahoma and on Block Island off the coast of Rhode Island. The reasons for this decline are thought to be habitat destruction and the loss of food sources. One theory is that the passenger pigeon, *Ectopistes migratorius*, which became extinct in 1914, was a major food source for the beetle; it started to decline at about the same time as the pigeon. In 1989 it was placed on the state and federal endangered species list. Today the American burying beetle is bred in captivity at the Roger Williams Zoo on Rhode Island. A new wild colony has been reintroduced onto the island of Nantucket off the coast of Massachusetts

BUSHBABY

THERE ARE NINE SPECIES of bushbabies, all of which live in sub-Saharan Africa. The classification of these closely related nocturnal primates is a subject of some dispute among scientists. The currently accepted view is that there are four genera of bushbabies, though at one time scientists classified all nine species together in the genus *Galago*.

The most widespread and best-known species is the Senegal bushbaby, *G. senegalensis*, which is 4¾–8 inches (12–20 cm) long; the tail, which is longer than the body, is bushy except at the base. The Senegal bushbaby has a round head, short muzzle and large eyes. The ears are large and flesh-colored and can be folded at will. The hind legs are long, the forelegs shorter, and the fur ranges from yellowish gray to brown, with a variable amount of white on the muzzle and throat. Like the only other species in its genus, the South African bushbaby, *G. moholi*, it is referred to as both a bushbaby and a galago.

Other closely related species include Demidoff's dwarf galago, *Galagoides demidoff*, which grows up to 13 inches (33 cm) long including its 8-inch (20 cm) tail. The thick-tailed greater bushbaby, *Otolemur crassicaudatus*, from East and South Africa, is 30 inches (76 cm) long; half of this length consists of its tail.

Behavioral patterns

Bushbabies live in groups in dry scrub, light woodland and bush; some species occur in dense forest. They belong to the suborder Prosimii, which includes lemurs, and are related to monkeys and apes. In common with many large primates, bushbabies regularly lick and groom one another's fur. During the day bushbabies rest together in dense foliage, a hollow tree, the fork of a branch or an abandoned bird's nest. At night, when they come out to feed, they move about singly or in pairs, their activity and agility contrasting with their slow movements if disturbed during the day. Bushbabies can take long leaps from branch to branch, with great skill and balance, covering as much as 15 feet (4.5 m) in one movement. On a level surface bushbabies sometimes run on the hind legs only.

Bushbabies use their hands to pick up food and objects for examination. Insects, especially locusts, form a major part of the bushbabies' diet, but they will also take flowers, pollen, honey, seeds and fruits, as well as birds' eggs and unfledged nestlings. Acacia gum is an important foodstuff for some species, including *O. crassicaudatus* and the silver galago, *O. argentatus*.

Communication and br   ng

Bushbabies show aggressio   rising on their hind legs and baring their   . When greeting each other they utter rasp   alls and use high and low notes to esta   and advertise a territory. They also m͵   ͵ritory by wetting their feet and hand͵   ͵ urine, saliva or secretions from the g͵   glands and smearing a scent onto branch   cientists believe that bushbabies make si   ͵se of wet paw marks to mark a trail by ·   they can find their way back after nocturr   aging.

The characte   cry of those species in the genus *Otolemu͵*   ͵milar to that of a human baby and ga·   ͵e to the name bushbaby. However, bu͵   ͵es employ a variety of other calls for dif   ͵ occasions. One analysis of bushbaby c   ͵as identified cackles and clicks for drawir   ͵ntion, grunts and squeaks when apprehen   ͵eezes when exploring, moans to show di͵   ͵and shrieks when alarmed. Bushbabies ͵   ͵ploy unique vocal communication signals͵   ͵vent interbreeding between species.

A bushbaby's huge ears are batlike in their mobility and sensitivity. They enable it to locate predators and prey in conditions of poor visibility, such as dense forest undergrowth.

Bushbabies are nocturnal and hunt by sight, so need large eyes for maximizing light. A reflective layer in the eye called the tapetum lucidum greatly improves light-gathering and glows brilliantly in flashlight.

Bushbabies have two breeding periods per year with two young, rarely one or three, born in each. The young are born from April to November after a gestation of about 4 months. The mother carries her young in her mouth, by the nape of the neck.

Young bushbabies can walk on all fours immediately after birth, stand on their hind legs after 24 hours and take short leaps within a week. They begin taking solid food in 2–3 weeks, and are fully weaned at 7–10 weeks. The life span of *G. senegalensis* is up to 16 years in captivity, but is probably much less in the wild.

Bushbaby pollination

Flowers are usually pollinated by insects, but there are instances of larger animals performing this task. Bushbabies often feed from the baobab, an African tree with large white flowers 5 inches (12.5 cm) across. These open after sunset to display many anthers, which contain the male pollen grains, the flower's fertilizing parts.

During studies carried out in East Africa scientists found that bushbabies of the genus *Otolemur* visited baobab trees on eight consecutive nights. The animals fed on the newly opened flowers, moving from one to another and burying their faces in them. When the flowers fell the next day, each bore signs of having been licked and chewed, and the small fleshy sepals had been almost entirely eaten. On subsequent nights a pale ring of pollen was visible around each bushbaby's face. However, the bushbabies had eaten only the outer parts of the flowers. The pistils, containing the ovary, style and stamen, were undamaged. By taking pollen from flower to flower without damaging the pistils, the bushbabies acted as agents of fertilization.

BUSHBABIES

CLASS **Mammalia**

ORDER **Primates**

FAMILY **Lorisidae (alternatively Galagonidae)**

GENUS **Bushbabies or galagos, *Galago* (detailed below); needle-clawed bushbabies, *Euoticus*; greater bushbabies, *Otolemur*; dwarf galagos, *Galagoides***

SPECIES **9, including Senegal bushbaby *Galago senegalensis*; South African bushbaby, *G. moholi*; thick-tailed greater bushbaby, *Otolemur crassicaudatus*; and Demidoff's dwarf galago, *Galagoides demidoff***

WEIGHT
⅓–⅔ lb. (125–300 g)

LENGTH
Head and body: 4¾–8 in. (12–20 cm); tail: 7–12 in. (18–30 cm)

DISTINCTIVE FEATURES
Very large eyes; large, mobile ears; long fingers with bulbous tips; long, bushy tail

DIET
Mainly acacia gum and large insects, especially locusts; also flowers, fruits, seeds, honey, bird eggs and nestlings

BREEDING
Age at first breeding: 10 months; breeding season: varies according to region; gestation period: 120–126 days; number of young: usually 2; breeding interval: 6 months

LIFE SPAN
Up to 16 years or more in captivity

HABITAT
Open woodland, scrub and wooded savanna

DISTRIBUTION
Sub-Saharan Africa

STATUS
Common

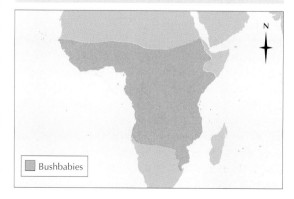

Bushbabies

BUSHBUCK

THE BUSHBUCK IS THE smallest antelope in the genus *Tragelaphus*, to which the nyala (*T. angasi*), kudu (*T. imberbis*) and sitatunga (*T. speki*) also belong. Males stand only 25½–43 inches (65–110 cm) at the shoulder and weigh a maximum of 170 pounds (77 kg). The color of the back and flanks ranges from light tawny in females to dark brown in males; the males are usually darker underneath. There is a considerable amount of white on the body, including patches on the throat, two spots on the cheeks below the eyes and traces around the lips and chin. White stripes run down the insides of the legs, and on the body there is a variety of spots and stripes running vertically and horizontally. The pattern of the markings varies greatly among the 20 or more subspecies.

Male bushbuck are larger than females, with a bushy mane along the length of the back that can be erected in fear or alarm, and a pair of sharp horns that may reach 22 inches (56 cm) in length. Females only occasionally bear horns. The spiral horns twist in slightly more than one complete turn and have keels (curves) on both front and back. Old males often develop patches of hairless skin around the neck. These marks are made by the tips of the horns when the bushbuck's head is thrown back as the animal makes its way through the dense cover that characterizes its natural habitat.

Shy and largely solitary

Bushbuck range across sub-Saharan Africa, from the shores of the Red Sea in Somalia and Ethiopia westward across the southern borders of the Sahara Desert to Senegal, and south to the Cape Province of South Africa. They favor swamps, woodland and bush, although they are found in most environments in Africa apart from open plains, deserts and ground over 10,000 feet (3,000 m) above sea level.

Bushbuck need to drink regularly and stay close enough to a water source to be able to drink at least once a day; they occasionally live in reed beds. If deprived of any alternative source of water, bushbuck can subsist on dew. They are

Female bushbuck are born with a reddish coat that becomes darker with age. Unlike males, females rarely grow horns.

BUSHBUCK

CLASS	**Mammalia**
ORDER	**Artiodactyla**
FAMILY	**Bovidae**
GENUS AND SPECIES	***Tragelaphus scriptus***

WEIGHT
Male: 66–170 lb. (30–77 kg);
female: 53–92 lb. (24–42 kg)

LENGTH
Male, head and body: 41–59 in.
(105–150 cm); shoulder height: 25½–43 in.
(65–110 cm). Female smaller than male.

DISTINCTIVE FEATURES
Spiral horns (usually absent in female) with
pronounced keels at front and back; variety
of white markings on body and legs

DIET
Leaves, twigs, fruits and grasses

BREEDING
Age at first breeding: 11–12 months;
breeding season: all year; gestation period:
about 180 days; number of young: 1;
breeding interval: 8 months

LIFE SPAN
Not known

HABITAT
Swamps and dense vegetation near water

DISTRIBUTION
Sub-Saharan Africa

STATUS
Common

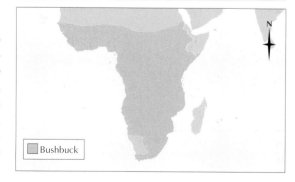

Bushbuck

There is considerable variation in bushbuck markings. Western forest bushbuck bear both horizontal and vertical body stripes and spots on their haunches. Bushbuck found in the more open woodlands to the east and south of the species' range are more sparsely marked and may even be plain.

mainly browsers, feeding on leaves, the tips of twigs and wild fruits such as acacia pods and wild olives; they graze infrequently.

Unlike most antelope, bushbuck lead solitary lives other than during the rut and breeding season when buck and doe or doe and calf will be seen together. Calves are born year-round, after a gestation of about 6 months, but most are born from mid-October to mid-November. Bushbuck are not seen often as they tend to emerge at dusk, although they bask in the early morning sun. They have very sensitive hearing and eyesight, and in areas where they are often disturbed skillful stalking is needed to observe them. Bushbuck are adept at wending their way through dense undergrowth, making use of low "tunnels" created by the passage of animals, and can make considerable leaps. They will take to the water when pursued.

Difficult to catch

Bushbuck bark, rather like a dog, when threatened or alarmed. Their main predators are leopards and Cape or African hunting dogs. Lions are not a serious threat because they hunt in open country rather than the woodland and swamps which the bushbuck prefer. Crocodiles have been reported to kill bushbuck by dragging them under as they drink at the water's edge, and also venture onto land to catch them as they feed.

Although considerably smaller than their relatives in the genus *Tragelaphus*, bushbuck can be aggressive, particularly when they are wounded. Predators do not find bushbuck easy prey, for they defend themselves vigorously and have been known to kill both leopards and Cape hunting dogs, usually impaling them on their horns. Their loud, roaring bark also serves to intimidate

predators and to alert other bushbuck to danger. The ability to move rapidly through undergrowth and to "freeze" when necessary also makes bushbuck difficult to track. Their variable marking breaks up the outline of the body and enables the bushbuck to effectively blend in with natural cover.

BUSHDOG

THE BUSHDOG IS A MEMBER of the family Canidae, which also includes wolves, jackals, foxes, coyotes and the domestic dog. Its fur is short and mainly reddish brown, with the head, neck and forequarters orange, tawny or white, and the belly, hindquarters and tail dark brown or almost black. The bushdog has an almost brushlike tail which grows to a length of 4¾–6 in. (12–15 cm); overall the animal grows to 22½–29½ in. (57–75 cm), head to tail. There are sometimes lighter patches on various parts of the body, including the throat and the underside. Stocky in build, the bushdog is a skilled swimmer and diver. Unusually for a canid, it has very short legs and broad feet which are webbed and serve as paddles. It is also able to swim underwater if necessary. The bushdog's tendency to hunt near or in water is a further distinction from other canids.

Bushdogs are found in northeastern and central South America, and range from Panama south to northeastern Argentina and west to the Pacific coast of Columbia and Ecuador. They are absent from the Andes. The species was initially described from subfossil bones which were found in Brazilian caves.

Seldom seen

The bushdog's natural habitat is woodland, savanna, forest and damp, grassy plains where the soil is sandy; however, it is very rarely seen. This may explain the former erroneous belief that the bushdog is nocturnal; in fact, it is a diurnal species that retires to a den at night. Bushdog dens are normally located in the bases of hollow trees and in burrows. A bushdog may make its own burrows, or utilize those abandoned by other animals, such as armadillos.

The bushdog's short legs, webbed feet and strong tendency to hunt near water make it unique in the dog family, Canidae.

In common with many other canids, bushdogs mark out their territories using urine and by spraying strong smelling liquids from anal glands. Vocalizations are also important in this respect. The calls that bushdogs make, including whines, squeals and doglike barks are similar to those made by the red fox.

Due to the rarity of sightings, little is known of the bushdog's reproductive habits, though it is believed to breed all year round. Litters generally consist of between one and six young and are usually born during the rainy season. Females are able to breed at 10 months of age and can give birth to two litters per year. While the females are nursing, the males bring food to them in the den. It is not known whether any other members of the pack, apart from the female's mate, help her rear the young.

Social hunter

Together with the Cape hunting dog and the Asiatic dhole, the bushdog is probably the most social member of the family Canidae and the most dedicated pack hunter. Although occasionally a solitary hunter, it generally hunts in packs of up to 10 individuals. There is some evidence that these packs may be family groups. Scientists believe that the bushdog will eat any animal flesh, including carrion, though it is also known to feed on some soft fruits and plant materials.

Large animals that live near water feature prominently in the diet of bushdogs, particularly pacas, agoutis and rheas. Bushdogs are also known to feed on the largest rodents of all, capybaras, which are eight to nine times larger than a bushdog. There are even reports of bushdogs attacking deer. Packs of bushdogs pursue prey into water and work together to ensure that the victim is caught. Some pack members chase the prey through the water, while others line the bank to effect a kill should the animal escape its swimming pursuers.

Vulnerable to change

Bushdogs are protected from trade and are rare throughout their range, though the reasons for their scarcity are not clear since they are not currently hunted for their fur or heavily persecuted as pests.

Scientists believe that the most likely cause of the decline in bushdog populations is the alteration or destruction of the species' natural habitat by humans. The widespread clearance of South America's rain forests, the bushdog's preferred habitat, is believed to have had the most damaging effect on the species. Where forests still stand, human settlement and the resulting pollution of streams and rivers have also had an adverse effect on bushdogs.

BUSHDOG

CLASS	**Mammalia**
ORDER	**Carnivora**
FAMILY	**Canidae**
GENUS AND SPECIES	*Speothos venaticus*

WEIGHT
11–15½ lb. (5–7 kg)

LENGTH
Head and body: 22½–29½ in. (57–75 cm); shoulder height: 12 in. (30 cm); tail: 4¾–6 in. (12–15 cm)

DISTINCTIVE FEATURES
Stocky, low-slung body; broad face; small ears; short legs and webbed feet; short, brushlike tail; reddish brown coat with white markings, darker on belly and tail and toward hind quarters

DIET
Large rodents, including agoutis, pacas and (rarely) capybaras; rheas and other large birds; some carrion and plant material

BREEDING
Age at first breeding: 10 months; breeding season: all year; gestation period: about 67 days; number of young: usually 2 or 3; breeding interval: about 240 days

LIFE SPAN
Up to 10 years in captivity

HABITAT
Forests, savanna and damp, grassy plains, often near water

DISTRIBUTION
Northern and central South America from Panama south to northeastern Argentina; absent from Andes Mountains

STATUS
Vulnerable. Little population data available due to rarity of sightings; appears to be widespread but scarce throughout range.

Bushdog

BUSHPIG

THE BUSHPIG RANGES ACROSS Africa south of the Sahara and is also found in Madagascar. It is stout-bodied, with a coat of short bristles and with longer, wiry bristles along the midline of the back and neck, on the sides of the face and the lower flanks. The head is disproportionately large, the muzzle is narrow and the tufted ears run to a point. A pair of bony protuberances is present on the face below the eyes; these are larger in old individuals but are usually inconspicuous because of the long hairs on the face. The bushpig's color varies.

Young adults are reddish but this coloring changes with age to a russet brown or black, particularly on the legs and shoulders. The long bristles of the face and the crest on the back are a mixture of white and black, the white predominating in some individuals, the black in others. The face often features heavy white marking.

The bushpig's tusks are not obvious while the mouth is shut. The upper tusk is 3 inches (7.5 cm) long, while the longer lower tusk measures about 7½ inches (19 cm). The upper tusk rubs on the lower one, resulting in a honing action which keeps both tusks sharp.

Bushpigs typically inhabit broken country with patches of forest or thicket, well-wooded rocky ravines, reed beds and readily available drinking water. The animals regularly roll in mud or in grass to cool down, and will rub their flanks against specific trees, gashing the bark with their tusks. Scientists believe that this may serve to mark out territory.

Sounders and social behavior

Bushpigs are highly secretive animals and rarely reveal themselves to humans. Mainly nocturnal, they come out by day only when undisturbed and will rapidly make for cover if they sense danger. They live in groups known as sounders, made up of an old boar, two or more sows and young of varying ages. One of the younger adult boars exercises control when the old boar is away. Groups of European wild boars have similar social structure, with a dominant boar and a subordinate animal, called a page. A sounder may consist of up to 20 bushpigs, though larger and more temporary associations of up to 60 animals also occasionally form.

Bushpigs feed chiefly on roots and tubers, invertebrates and wild fruits but, like domestic pigs, readily eat flesh when available, including eggs and young birds. To obtain the fruits from small trees, bushpigs lean heavily against the trunks to topple them. Bushpigs run fast, swim well and aggressively defend themselves against intruders. The first signs of aggression in bushpigs include the raising of the bristles on the back. Because of the pigs' strength and sharp tusks, predators are few, and probably include only lions and leopards, which may kill the young. Leopards are likely to be the chief predator, but there are reports of leopards being forced to take cover in trees by the hostile actions of adult bushpigs.

The main breeding season for bushpigs is from March to August and the gestation period is about 4 months. As a result the young are born

The position of a bushpig's ears displays the animal's intentions. Vertically raised ears indicate submission; horizontal ears are a sign of aggression.

Bushpigs show a preference for river courses and swamp regions and regularly wallow in mud.

from July to January, usually during the rainy season. The sow makes a bower in long grass for her litter. The young, when first born, are brown with yellowish longitudinal stripes, which later become rufous brown. A litter includes three or four young, occasionally eight, although the sow has only three pairs of teats. Bushpigs have lived as long as 20 years in captivity.

Destructive cultivators

Many African farmers regard bushpigs as a menace to cultivated crops. As well as eating crop plants and trampling them underfoot, bushpigs cause damage by rooting with their snouts to search for food and by plowing through the crops. Moreover, bushpigs are robust and heavy animals, and can be aggressive when attacked; these qualities combined with their razor-sharp teeth are enough to drive away farmers' dogs.

It has been suggested that it is precisely those features of the bushpig's behavior that infuriate farmers that make it a valuable cultivator in wild country. The bushpig plows the land, especially during the rainy season, loosening and aerating the soil. It speeds up the decomposition of dead branches and twigs by breaking them up with its hooves and trampling them into the churned-up soil, fertilizing the ground and dispersing seeds with its droppings. Some trees are completely dependent on the bushpig's cultivations because they bear fruit underground out of reach from most insects, birds and mammals. The bushpig digs out the fruit with its snout, and eats and then scatters the undigested seeds over the land.

BUSHPIG

CLASS	**Mammalia**
ORDER	**Artiodactyla**
FAMILY	**Suidae**
GENUS AND SPECIES	***Potamochoerus porcus***

ALTERNATIVE NAMES
African bush pig; red river hog

WEIGHT
100–285 lb. (45–130 kg)

LENGTH
Head and body: 39½–59 in. (100–150 cm); shoulder height: 23–38 in. (58–96 cm); tail: 12–17 in. (30–43 cm)

DISTINCTIVE FEATURES
Long, pointed ears, with tufts at tips; light-colored mane along neck and midline of back; long bristles on face and lower flanks; coat reddish, rich brown or black according to age

DIET
Roots, tubers, invertebrates and fruits; also bird eggs, nestlings and small vertebrates

BREEDING
Age at first breeding: 3 years (female); breeding season: March–August; gestation period: 120 days; number of young: usually 3 or 4; breeding interval: about 1 year

LIFE SPAN
Up to 20 years in captivity

HABITAT
Forests; riverine and montane habitats with dense vegetation

DISTRIBUTION
Sub-Saharan Africa except Namibia and most of Botswana and South Africa; also Madagascar and Mayotte Island in the Comoros Archipelago

STATUS
Fairly common

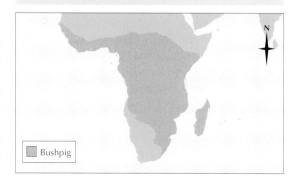

Bushpig

back and inflates from its throat a great fold of feather-covered skin, which hangs down to the ground. At the same time it utters a far-carrying booming note that sounds from a distance almost like a roaring lion.

The male great bustard undergoes a remarkable transformation when it displays. From being a relatively dull, gray and brown bird it suddenly becomes a billowing mass of white feathers. The feathers on the back are turned over, and the tail is turned up and over the back, to display their white undersides. At the same time, the head sinks onto the shoulders, the mustachelike bristly, white feathers that sprout from the cheeks stand upright and the air sac under the throat is inflated. As the legs are bent and the wings fluffed out, the bustard looks like an untidy ball of feathers.

Bustards make little in the way of a nest, merely trampling a patch in the grass or scraping a depression in the ground. The female incubates the eggs. Sometimes the male stands nearby, but some species are promiscuous, the male and female going their separate ways after mating. The eggs, from 1 to 5 in number according to the species, hatch in about 1 month and the chicks can leave the nest and run about as soon as they are dry. The female looks after them for another month or more, at first catching insects to give them in her bill. Later the chicks catch their own insects, but they stay with their mother until they are able to fly.

The human menace

Despite their size and exceptional wariness bustards fall victim to many predators, as do the majority of ground-nesting birds. Foxes kill both adults and young, and crows take eggs. As so often happens, however, humans are the birds' worst enemy. A bustard the size of a chicken or turkey makes a good meal, and bustards are regularly hunted for the pot. This has led to their elimination from many former haunts. The houbara bustard, *Chlamydotis undulata*, in particular has suffered from excessive hunting, because it is a traditional quarry of Arab falconers in the Middle East. Many bustard species are now protected by law, but the combination of intensive agriculture, illegal hunting and urbanization may yet push some of them to extinction.

Male great bustards perform spectacular courtship displays that can be seen from afar in the flat landscapes the species inhabits.

BUTCHERBIRD

To human ears the flutelike, mellow piping of the gray butcherbird (above) and pied butcherbird are among the finest of all bird songs.

THE SIX BUTCHERBIRDS OF Australia and New Guinea, together with the currawongs and bell-magpies, also of Australia, make up the family Cracticidae. The name butcherbird refers to the habit of impaling surplus prey on thorns as a food store. Another bird that shares this behavior, the red-backed shrike, *Lanius collurio*, of Eurasia and Africa, was once sometimes called the butcherbird. The shrike's nickname has fallen into disuse, however, and the name butcherbird now refers exclusively to the six Australasian species.

Butcherbirds are stocky birds, about 10–23 inches (25–58 cm) long, with large heads and powerful, hooked bills. Their plumage is primarily black-and-white. As its name suggests, the gray butcherbird, *Cracticus torquatus*, also has gray in its plumage, with black on the face, crown, nape, wings and tail. Females are generally more brown on the crown and upperparts. Another species, the pied butcherbird, *C. nigrogularis*, has a glossy black head, neck and back, and white wings.

Butcherbirds are found in light woodland and open, scrubby country with scattered trees. Two species are restricted to New Guinea: the black-headed butcherbird, *C. cassicus*, and the white-rumped butcherbird, *C. louisiadensis*. Two other species, the black butcherbird, *C. quoyi*, and the black-backed butcherbird, *C. mentalis*, occur in both New Guinea and northern Australia. The gray butcherbird ranges over most of Australia and Tasmania. It prefers open forests and scrub around the coasts and thickly timbered mountain ranges, but is also found inland, in the woods that are scattered over the plains. The pied butcherbird is also widespread, but nowhere common. It is usually encountered in pairs, for butcherbirds stay paired year-round, in pine scrub and timber clumps on the plains.

Fine singers

Butcherbirds are shy by nature but become tame near human dwellings. Here they are welcomed for their lilting, cheerful song, composed of long phrases repeated in different orders. The song bears no relation to the song of any European bird, except perhaps to the liquid phrases of the European blackbird.

The pied butcherbird is considered by many to be one of the avian world's best singers. It was once described as chanting, slowly and richly, "this is the tree," but this does not adequately convey the beauty of its song. Male and female sing equally well and regularly perform duets. The song can be heard year-round, often on moonlit nights. For this reason the species was traditionally called the break o' day boy. In fall and winter, the pied butcherbird's song changes and is less sustained. In common with the black-backed butcherbird, it also mimics the songs and calls of other, unrelated birds.

The term songbird is usually limited to the suborder Oscines of the order Passeriformes. The Oscines includes the familiar songbirds as well as bulbuls, mockingbirds and babblers. It also contains less likely candidates, among them butcherbirds, crows, shrikes and nuthatches. The basis for this distinction lies in the anatomy of the syrinx (voice box), which has the same function as the larynx of mammals. The syrinx lies in the windpipe and has a resonating chamber and vocal chords, the tension of which is altered by muscles to create different noises. The suborder Oscines differs from other birds in having a more complex system of control, involving five to nine pairs of muscles, thus allowing a wide variation in the sounds that can be produced.

GRAY BUTCHERBIRD

CLASS	**Aves**
ORDER	**Passeriformes**
FAMILY	**Cracticidae**
GENUS AND SPECIES	***Cracticus torquatus***

ALTERNATIVE NAMES
Silver-backed butcherbird; Derwent jackass; Tasmanian jackass; whistling jack

LENGTH
Head to tail: 9½–12 in. (24–30 cm)

DISTINCTIVE FEATURES
Finely hooked bill; stocky body. Male: gray with black crown and cheeks and partial white collar. Female: browner plumage.

DIET
Small lizards, snakes, birds and mammals; large insects such as beetles and crickets

BREEDING
Age at first breeding: 1 year; breeding season: July–March; number of eggs: 3 to 5; incubation period: about 23 days; fledging period: 25 days; breeding interval: 1 year

LIFE SPAN
Not known

HABITAT
Rain forest margins, eucalypt woodland, scrub, parks and gardens

DISTRIBUTION
Much of Australia, except for arid interior and rain forests of northeast; also Tasmania

STATUS
Fairly common

Gray butcherbird

Butcherbirds are extremely rapacious predators for their size. They hunt large insects, such as beetles, crickets and grasshoppers, as well as small lizards and snakes, field mice and a few small birds. Butcherbirds take up position on a favored lookout post, such as a dead branch,

fence or telephone wire, from which they scan the ground for passing prey. They then pounce while the prey is on the ground.

Surplus food is stored by impaling it on thorns and barbed wire, or in some cases by wedging corpses into a tree crotch, between the trunk and branch. Butcherbirds have also been seen to drag fledgling songbirds along the jagged end of a broken branch so that the prey lodged there while the butcherbirds ripped off the flesh. They can prove to be a nuisance to bird fanciers, grabbing caged birds and dragging them out through the mesh of the enclosure.

There have been many observations of butcherbirds hunting in unison with the little falcon, *Falco longipennis*, also known as the Australian hobby. A butcherbird takes up position near to one of the falcons. Small animals are constantly being disturbed by the presence and hunting activities of the falcon, and the butcherbird swoops on these as they dash for cover.

Strongly territorial

Both male and female butcherbirds guard their territory by singing and chasing intruders. The territory may be defended year-round, but during the breeding season defense becomes vigorous, and predators such as snakes and ravens are attacked. The year-round pairing of butcherbirds contrasts with the behavior of their near-relatives the bell-magpies, in which clans of up to 20 birds defend a large, communal territory.

A butcherbird nest is an untidy structure of twigs, lined with fibrous roots and grasses. It is about 9 inches (23 cm) in diameter, with a cup 2 inches (5 cm) deep, and is placed in a tree above the ground. Sometimes the nest is so thinly built that the eggs can be seen through the base. Eggs usually number three to five. In dry areas, the pied butcherbird lays a single egg. The chicks fledge about 25 days after hatching.

The pied butcherbird, like all the members of the genus Cracticus, *caches surplus food on thorns and inside crevices to eat later.*

BUTTERFISH

Butterfish live at the margin of sea and land and so all activity is dominated by the tides. When the water drains away at low tide, the fish seek refuge under stones and in seaweed.

THE BUTTERFISH IS AN eel-like species of blenny living between the tidemarks on North Atlantic coasts, as far south as Wood's Hole, Massachusetts, in the west and the English Channel in the east. At one time it was called nine-eyes, after the row of dark spots set along the base of the dorsal fin. The butterfish is ribbon-shaped, with a low dorsal fin running the length of the back, a long, low anal fin, a small tail and very small paired fins.

The name butterfish refers to the slippery nature of this fish and to the difficulty humans have in holding it between the fingers. Since this is a characteristic of many fish, it is not surprising to find a New Zealand wrasse also called a butterfish, as well as two other unrelated species in coastal waters of the United States, all of which have slimy bodies.

Uncovered at low tide

The butterfish feeds on small, bottom-living invertebrates, and is diurnal (day-active). Like other blennies, it is normally seen when the tide is out, sheltering under stones and seaweed, or in damp rock crevices. It becomes more active when, at high tide, it is completely covered with water. As with other blennies, butterfish living close inshore alternate active periods with resting periods throughout the daylight hours, corresponding to the rise and fall of the tide.

Spawning occurs from January to February. The female lays her eggs in a mass, usually in a cavity in a rock or in an empty bivalve mollusk shell. The female curves her body into a loop while laying, so rolling the egg mass into a ball. Once the mass is laid the male coils around it and continues to protect the eggs in this way until they hatch, a month later. The larvae are pelagic, drifting out to sea and spending several months

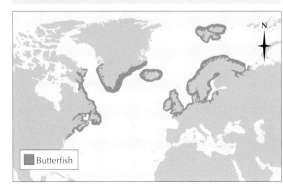

Butterfish

at depths of about 200 feet (60 m) before returning to shallow coastal waters. Egg-clustering in the butterfish ensures that all the eggs benefit from the protection of the male. Individual eggs would be impossible for one or two parents to guard and would soon be lost to predators.

BUTTERFLIES AND MOTHS

MORE THAN 160,000 SPECIES of butterflies and moths have been described so far, and many new species are discovered every year, especially in the Tropics. Butterflies and moths occur worldwide and belong to the order Lepidoptera. Zoologists formerly divided this large and well-studied order of insects into two groups: the Rhopalocera (butterflies) and the Heterocera (moths). Any separation of moths from butterflies is largely artificial, however, because their structural and behavioral differences are highly complex and far from clear-cut.

Scientists now place members of the Lepidopera in five or more suborders, depending on which classification system is used. The large and diverse suborder Ditrysia contains all butterflies and most moths, including several families of moths that are more closely related to butterflies than they are to other moths. The remaining, relatively primitive, species of moths are contained in the suborders Zeugloptera, Exoporia, Dacnonypha and Monotrysia, although some authorities classify these primitive moths in up to 12 suborders.

Structure

Being insects, all lepidopterans have three well-defined body segments, the head, thorax and abdomen, and three pairs of legs, although in some families the front two legs are reduced to stumps. Adult lepidopterans typically have two pairs of membranous wings covered with thousands of overlapping scales; large compound eyes; long, multisegmented antennae; and a tubular structure called a proboscis for sucking nectar

Some butterflies and moths, including the atlas moth, **Attacus atlas,** *are larger than small birds, while at the other end of the scale are species measuring just a few millimeters.*

or other liquid foods. The larvae usually have an armored head; strong mandibles for chewing a variety of foods; three pairs of legs on the thorax; and several pairs of false legs on the abdomen, which is divided into 10 segments. However, numerous species depart from this general description. For example, in adult clearwing moths (Sesiidae) and butterflies

CLASSIFICATION Butterflies and Advanced Moths	CLASSIFICATION Primitive Moths
CLASS Insecta	**CLASS** Insecta
ORDER Lepidoptera	**ORDER** Lepidoptera
SUBORDER Ditrysia	**SUBORDER** Zeugloptera, Exoporia Dacnonypha, Monotrysia
NUMBER OF SPECIES Around 155,000	**NUMBER OF SPECIES** Around 5,000

The European swallowtail, Papilio machaon, *is typical of the true butterflies. It has prominent clubs at the ends of its antennae and when perched holds its colorful wings upright.*

Size matters

Lepidopterans vary in size from the huge to the minute. Largest of all are the Queen Alexandra's birdwing butterfly, *Ornithoptera alexandrae*, and the hercules moth, *Coscinoscera hercules*. Both species, which inhabit tropical rain forests in New Guinea, can attain wingspans of 11 inches (28 cm). Of similar size are the atlas moth of Southeast Asia and the owlet moth, *Thysania agrippina*, which occurs from the southern United States south to South America. The smallest lepidopteran known to science is a micromoth, *Stigmella ridiculosa*, of the Canary Islands in the northeastern Atlantic. This moth has a miniscule wingspan of about 2 millimeters. Certain blue butterflies (Brephidium) from North America and South Africa measure ½ inch (1.5 cm) across the wings.

Coloration

Adult lepidopterans obtain their tremendous range of colors from the rows of scales covering their wings and bodies. The scales contain pigments or have finely sculptured surfaces that appear metallic or prismatic as a result of light diffraction. For many butterflies and moths color is vital to survival. It can serve as camouflage, enabling the insects to blend in with their surroundings and evade predators. Many species have evolved markings that resemble leaves or bark. Several of the kallima butterflies found in southern and Southeast Asia bear an extraordinary resemblance to dead leaves. Not only are the wings colored in shades of brown, they are actually shaped like leaves and are traced with "veins" like those of leaves.

Alternatively, color can make butterflies and moths stand out and attract attention, and it is therefore one of the main methods of attracting potential mates. Among the more spectacular butterfly families are the morphos (Morphidae), birdwings and swallowtails (Papilionidae) and metalmarks (Riodinidae). Moths with typically dramatic coloration include the tigers (Arctiidae), emperors (Saturniidae), hawkmoths (Sphingidae) and uranids (Uraniidae). The wings of most butterflies have relatively bright uppersides and less striking, softer undersides. This enables the butterflies to perch inconspicuously by closing the wings back to back, and give an impressive display of color by opening the wings again.

(Satyrinae) large areas of the wings lack scales, revealing the transparent underlying membrane. The superfamily Micropterigoidea contains about 100 species of primitive moths that, as adults, have mandibles instead of a proboscis.

Butterfly or moth?

There are several general differences between butterflies and moths that hold for the majority of species. Moth antennae tend to be feathery or straight, whereas butterfly antennae are filamentous (thin) and clubbed (club-ended). The wings of moths are linked by structures that are missing in butterflies. Moths are often night-flying and dull in color, while butterflies are generally more brightly colored and always fly during the day. When at rest moths normally fold their wings to cover the body in a tentlike fashion; resting butterflies hold their wings vertically, above the body. However, there are many contradictions that make it impossible to define in absolute terms the differences between butterflies and moths. Burnet moths (Zygaenidae), for instance, have clubbed antennae, fly by day and are conspicuously colored.

Butterflies and Moths Family Tree

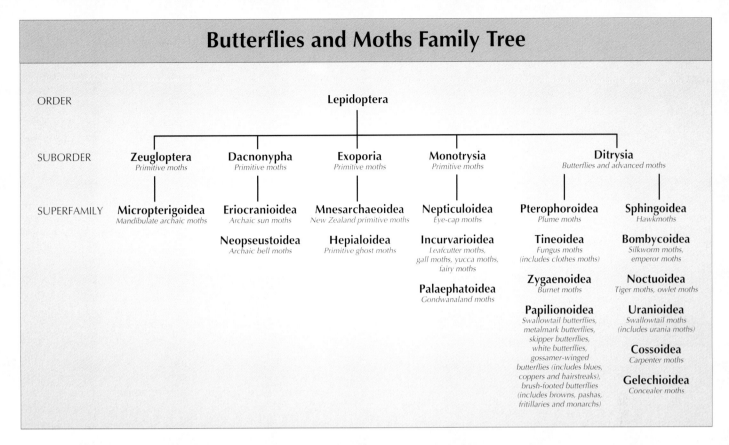

ORDER			Lepidoptera			
SUBORDER	**Zeugloptera** *Primitive moths*	**Dacnonypha** *Primitive moths*	**Exoporia** *Primitive moths*	**Monotrysia** *Primitive moths*	**Ditrysia** *Butterflies and advanced moths*	
SUPERFAMILY	**Micropterigoidea** *Mandibulate archaic moths*	**Eriocranioidea** *Archaic sun moths*	**Mnesarchaeoidea** *New Zealand primitive moths*	**Nepticuloidea** *Eye-cap moths*	**Pterophoroidea** *Plume moths*	**Sphingoidea** *Hawkmoths*
		Neopseustoidea *Archaic bell moths*	**Hepialoidea** *Primitive ghost moths*	**Incurvarioidea** *Leafcutter moths, gall moths, yucca moths, fairy moths*	**Tineoidea** *Fungus moths (includes clothes moths)*	**Bombycoidea** *Silkworm moths, emperor moths*
				Palaephatoidea *Gondwanaland moths*	**Zygaenoidea** *Burnet moths*	**Noctuoidea** *Tiger moths, owlet moths*
					Papilionoidea *Swallowtail butterflies, metalmark butterflies, skipper butterflies, white butterflies, gossamer-winged butterflies (includes blues, coppers and hairstreaks), brush-footed butterflies (includes browns, pashas, fritillaries and monarchs)*	**Uranioidea** *Swallowtail moths (includes urania moths)*
						Cossoidea *Carpenter moths*
						Gelechioidea *Concealer moths*

Color is also an important defense mechanism used by the caterpillars, or larvae, of butterflies and moths. For instance, many larvae that feed on grasses or on leaves in the tree canopy are green, but some species have evolved more sophisticated defensive coloration. The larvae of the comma butterfly, *Polygonia c-album*, closely resemble bird droppings. Larvae of the European puss moth, *Cerura vinula*, have black, eyelike markings on the head and prominent red threads at the end of the abdomen. When attacked, a puss moth larva rears up and lashes these threads from side to side, thus giving the impression of being a formidable opponent.

Chemical warfare

Plants often produce chemicals that make them repellant to herbivorous animals, especially the larvae of butterflies and moths. Likewise, numerous lepidopterans possess chemical weapons to deter birds and other predators. Some lepidopterans are poisonous in both the larval and adult stages; examples include burnet moths, tiger moths in the family Ctenuchidae and butterflies in the families Danaidae and Heliconiidae. The North American monarch, *Danaus plexippus*, or milkweed butterfly, overcomes the toxins in its food plants, absorbing the chemicals to make itself unpalatable.

Distasteful lepidopterans often advertise their unpleasantness through bright warning colors. Birds that eat such a species learn to associate unpleasant food with a particular color pattern and in future avoid individuals that have that appearance. Warning coloration is so effective that some edible butterflies and moths have evolved coloration that mimics genuinely distasteful species. For example, the viceroy butterfly, *Limenitis archippus*, is entirely harmless but looks like the unpleasant-tasting monarch.

From egg to adult

A butterfly or moth begins its life cycle when the adult female lays her fertilized eggs on or near a suitable source of food. The small eggs hatch into larvae that feed voraciously, regularly shedding their skin as they grow. Most lepidopteran larvae feed on plants but some eat wood, grain or wool and a handful of species are carnivorous. After a period of several days, weeks or months, depending on the species and on environmental conditions such as temperature, each larva changes

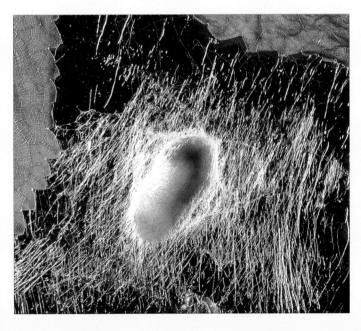

A larva of the silk moth, Bombyx mori, *inside its cocoon. Each larva, or silkworm, can produce several kilometers of silk.*

One of the best known migratory butterflies is the monarch. As temperatures plummet in the fall the species leaves North America and heads for Mexico, where vast swarms occur all winter.

The fastest-flying species is the death's head hawkmoth, *Acherontia atropos*, of Eurasia, which can accelerate to almost 34 miles per hour (54 km/h) for brief periods. Maintaining high airspeeds is very costly in terms of energy, however, and is most unusual among butterflies and moths.

Lepidopterans tend to use scents and sounds to find and court mates, although visual displays are important once partners have met. Powerful chemicals produced by some female moths can be detected by males at great distances. The male common emperor moth, *Eudia pavonia*, has perhaps the most acute sense of smell in nature: it can smell receptive females 6¾ miles (11 km) away. When they are with a potential mate, lepidopterans sample the air for traces of pheromones (chemical stimulants) to verify that they have encountered the right species. Having mated, the female flies away to lay her eggs. Some species, such as the Mediterranean brocade moth, *Spodoptera littoralis*, or cotton leafworm, lay thousands of eggs in batches. Others lay a few eggs, each in a separate place.

into a pupa, or chrysalis. The larva's head is equipped with a pair of spinnerets from which silk is extruded. This allows the pupa to produce a thread to attach itself to a twig or vegetation. Some larvae wrap themselves in leaves for added protection, others burrow underground, while the larvae of moths and a few butterflies spin a silken cocoon (sling) in which the pupa can transform into an adult. It takes several weeks or months for the metamorphosis to be complete, during which time the larval body is broken down and the tissues reassembled to create the adult form.

Life as an adult

When the adult butterfly or moth emerges, sometimes using chemicals to soften the pupal case, it pumps blood into its wings to stiffen them and before long it flies off. Most adult lepidopterans use the proboscis to suck water, nectar and other fluids from food sources such as fruits and flowers. The proboscis is coiled beneath the head when not in use and must be unrolled for feeding. It can be more than 8 inches (20 cm) long in some hawkmoths, which must reach nectar deep inside trumpet-shaped flowers. In other lepidopterans the proboscis may be very small. However, some species do not feed at all as adults.

It is rare for the adults of a typical butterfly or moth to live longer than 2–3 weeks; their sole purpose in life is to reproduce. Some species hibernate over the winter, so that the adult life span may last for 12 months. Others migrate toward the equator, where the winter is mild, before returning to cooler zones in the spring. In general, however, the adult stage of a lepidopteran is by far the shortest part of its life cycle.

Adult butterflies and moths are to a varying degree aerial. Many are powerful and agile fliers, while others fly weakly. The females of some moths, such as the gypsy and vaporer moths (Lymantriidae), lack wings and are therefore flightless.

Conservation

It is estimated that several hundred species and subspecies of butterflies and moths are now threatened or endangered. The main threats facing lepidopterans are habitat loss, pollution and collection by humans. Many species are tolerant of change in their environment and can successfully adapt to damaged habitats; however, large numbers depend on specific food plants and are very sensitive to chemical pesticides and poor air quality. For this reason scientists frequently use butterflies and moths as indicators of ecological change. If, in a given area, certain species are absent or in decline, this is often a sign that the habitat has suffered in some way.

For particular species see:
- APOLLO BUTTERFLY • ATLAS MOTH • BIRDWING
- BLUE BUTTERFLIES • BOLLWORM
- BROWN BUTTERFLIES • BURNET MOTH
- CLOTHES MOTH • COPPER BUTTERFLIES
- CORN BORER • FRITILLARIES • GOATMOTH
- HAIRSTREAK BUTTERFLIES • HAWKMOTH
- KALLIMA BUTTERFLIES • MONARCH • MORPHO
- OWLET MOTH • PEPPERED MOTH • PLUME MOTH
- PROCESSIONARY MOTH • PURPLE EMPEROR
- PUSS MOTH • SILK MOTH • SKIPPER
- SWALLOWTAIL • SWIFT MOTH • TIGER MOTH
- URANIA MOTH • VANESSA BUTTERFLIES
- WHITE BUTTERFLIES • YUCCA MOTH

BUTTERFLY FISH

IT IS DIFFICULT TO DISCUSS butterfly fish without confusion because the name is commonly used for different species of unrelated fish. The numerous species of marine angelfish that inhabit coral reefs are sometimes called butterfly fish after their colorful appearance. However, this article is concerned with a single freshwater fish, *Pantodon buchholzi*, the sole member of the family Pantodontidae. It is called the butterfly fish because of its large, winglike fins, which in the past have given rise to speculation that the species is able to fly.

Never more than 5 inches (12.5 cm) long, *P. buchholzi* lives in the rivers of tropical West Africa. It is colored gray green to brownish silver, and is marked with spots and streaks. The head and body are boat-shaped, flattened above and bluntly rounded below. The large mouth is directed upwards, and the nostrils are tubular. The most unusual feature of the fish is its fins. Each pelvic fin has four very long, filamentous

rays not connected to one another, and the unpaired fins are large, transparent and supported by long rays. The long ray fins are used in shallow water as stilts, on which the freshwater butterfly fish can support itself for long periods.

A flying fish?

The freshwater butterfly fish spends most of its time just below the surface of still or stagnant waters in the Congo and Niger basins, in weed-rich backwaters and standing pools. It is renowned for its ability to leap up to 6 feet (1.8 m) out of the water. However, scientists were for a long time uncertain whether the fish uses its large pectoral fins to glide in the air, as true flying fish do. The freshwater butterfly fish was formerly credited with flapping its fins in true powered flight in a manner reminiscent of bats and birds. It had been generally agreed that this was not the case when, in 1960, P. Greenwood and K. Thomson investigated the species' anatomy in detail.

The butterfly fish has the rare ability to prop itself as if on stilts, by means of the four long rays attached to each of its two pelvic fins.

FRESHWATER BUTTERFLY FISH

CLASS	**Osteichthyes**
ORDER	**Osteoglossiformes**
FAMILY	**Pantodontidae**
GENUS AND SPECIES	*Pantodon buchholzi*

LENGTH
4–5 in. (10–13 cm)

DISTINCTIVE FEATURES
Large, winglike pectoral fins; each pelvic fin has 4 long rays unconnected to one another

DIET
Small insects

BREEDING
Age at first breeding: probably 1 year; hatching period: 2–3 days

LIFE SPAN
Not known

HABITAT
Rivers, pools and streams

DISTRIBUTION
Watersheds of Niger and Congo Rivers

STATUS
Unknown; little population data available

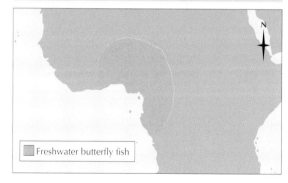

Freshwater butterfly fish

Exhaustive research has shown that butterfly fish are theoretically capable of flight above water, although it seems very unlikely that they do in fact fly.

They found it had a most unusual shoulder girdle, the arrangement of bones to which the pectoral fins are attached. In fact, the two authors described the girdle as unique among fish. The bones were so thin that the scientists had to be very careful not to damage them while dissecting specimens. The whole of the shoulder girdle is broad and flattened to give support to a highly developed system of muscles, which is comparable with even the greatly enlarged pectoral muscles that power the wings of birds.

The two scientists also found that the fins could not be folded against the body, as is usual in fish, but could be moved up and down. In brief, they concluded that, while it was still unproven whether or not the butterfly fish could make a powered flight, its shoulder girdle and muscles were such that it ought to be able to fly. However, there is not a single scientific study since 1981 that suggests the butterfly fish is capable of powered flight. The best that can be said is that the species has been seen to beat its fins up and down when held in the hand.

Feeding and breeding

The diet of the butterfly fish consists almost entirely of small insects, such as flies. These fall onto the surface of the water and are snapped up or are taken as the fish leaps from the water. The butterfly fish is largely nocturnal, emerging to feed at dusk when most insects are active. This behavior also helps it avoid the many day-active predators that share its habitat.

Relatively little is known about the breeding habits of the butterfly fish, and such details are drawn from the few that have been bred in captivity. Numerous false matings have been seen, with the male riding on the back of the female, holding her firmly with the long rays of the pelvic fins. Mating finally is effected by the two fish twisting their bodies to bring together the vents, located at the rear of the belly near the anal fin. It is therefore generally assumed that fertilization is internal. When the eggs are laid they float to the surface, adhering to the underside of floating aquatic plants. The fry remain at the surface, feeding on springtails, aphids and other tiny insects.

BUTTON QUAIL

BUTTON QUAILS ARE small birds, 4–7 inches (10–18 cm) long, resembling true quails both in their appearance and in their behavior. This group of 16 species is thought to be related to sandgrouse and pigeons. They have three toes on each foot, with no hind toe. The wings are rounded and the tail is extremely short. Button quails are sometimes called hemipodes, meaning "half a foot"; this is a reference to the birds' lack of a hind toe. The same name is frequently used to refer to the single European species.

Freezing in the grass

Button quails are secretive, generally ground-dwelling birds that hide among the undergrowth in pairs or small groups. They live on grass seeds, young shoots and insects that they scratch from the ground or find among the undergrowth. Button quails favor tropical or subtropical climates and usually make their home in dry bush or open country. Occasionally they are found in savannas, open woodland and marshy areas.

When they are frightened, button quails fly from their nests but only for a short distance. Thereafter they settle again in the grass where they are hard to locate. One notable exception in this respect is the lark quail, *Ortyxelos meiffrenii*, which flies strongly and may alight some distance from its starting point. When they have landed after a disturbance, button quails freeze until the danger appears to have passed and then slowly creep away through the vegetation, walking on the tips of their toes. This makes observation of their habits and attempts to find their nests a difficult task.

Although in general quails do not favor long-distance flying, a few species are partly migratory, including the yellow-legged button quail, *Turnix tanki*, and the common button quail, *T. sylvatica*. Button quails are found in many parts of the Old World. The common button quail ranges from Spain (where the species is known as the Andalusian hemipode), across Asia to the Philippines and throughout most of Africa, excluding the Sahara Desert and the tropical forests of the Congo basin. It is now virtually extinct in Spain, where there are currently no more than 10 pairs left. Other button quails live in Madagascar, Australia, New Guinea, Melanesia and Mindanao in the Philippines.

Role reversal

The roles of male and female button quails in courtship and rearing offspring runs counter to the behavior shown by many species. Females play a dominant part in courtship displays and, having laid their eggs, leave all incubation duties to the males. The changed roles of the sexes are reflected in their appearance. The male button quail generally has drab plumage, usually a combination of brown, gray and cream coloring. Its plumage provides it with some camouflage, which is important for a bird that has to sit as inconspicuously as possible on the nest. Female button quails are often larger and more brightly colored than males. The female barred bustard quail, *T. suscitator*, even acquires a courtship plumage in the breeding season.

During the breeding season the female calls to attract the male. The call of the common button quail is a booming sound, heard most frequently at dawn and dusk. In Spain the bird is known as the torillo, meaning "little bull," because of the resemblance of its call to the distant lowing of cattle. The female button quail has a special organ to produce this booming sound, formed by the inflation of the throat and windpipe, that acts as a resonator.

Having attracted a male, the female then proceeds to court him. She struts around in a circle with tail raised and chest puffed out, at the same time uttering her booming call, stamping and

Among button quails it is the males (above) rather than the females that take care of incubation.

pecking at the ground. Sometimes the female builds the nest on her own. At other times both sexes contribute to its construction, bringing grasses and dead leaves to the site and flicking them over their shoulders toward the growing pile of material.

There is much variation in the size and form of the finished nest, depending on the bird itself. In some species, such as the Australian black-breasted button quail, *T. melanogaster*, the nest is no more than a scantily lined depression in the ground. Other button quails form a roof over the nest by drawing stems of grass together to provide a dome.

The eggs are oval-shaped and there are usually four in a clutch. They are generally glossy, pale gray or buff-colored, and feature freckles of chestnut, dark brown or gray. After the eggs have been laid the female deserts the nest and leaves the male to incubate and rear them without her assistance. They hatch in 12–14 days, and almost immediately the young are able to follow their male parent about while he searches for food, which he presents to them in his bill. Button quail chicks acquire the ability to fly within 2 weeks but they stay together as a family for another 2 weeks.

It is very unusual for birds with such a short incubation time to be so well developed and active soon after hatching. Chicks spending less than 2 weeks in the egg are usually blind, naked and helpless when they hatch, though generally chicks attain the adult size and weight within about 6–7 weeks. Button quails also reach sexual maturity early, being able to breed in less than 1 year. Some female button quails bred in captivity may be capable of laying within just 3–5 months of their hatching date.

Aggressive females

Female button quails fight among themselves, as the males of other species do. In India the females of the barred bustard quail are trapped using decoys, models painted to look like other female barred bustard quails. The females caught in this manner are then set to fight against each other in the same way that male birds are in cockfighting.

The female button quail mates with several males. Having provided one male with a nest and eggs to look after, she then departs to mate with another. For the male to mate with several females, each having her own nest and clutch, is not unusual among birds, but for a female to have several clutches is surprising, because the difficulty of getting enough food to form the eggs usually limits the size of the clutch. It appears that female button quails have access to a very abundant food supply.

COMMON BUTTON QUAIL

CLASS **Aves**

ORDER **Gruiformes**

FAMILY **Turnicidae**

GENUS AND SPECIES *Turnix sylvatica*

ALTERNATIVE NAMES
Little button quail; little bustard quail; kurrichane button quail; Andalusian hemipode

WEIGHT
1–2 oz. (32–54 g)

LENGTH
Head to tail: 6–6¼ in. (15–16 cm); wingspan: 9¾–12¾ in. (25–30 cm)

DISTINCTIVE FEATURES
Rounded wings; almost tailless; orange rufous breast; bold black spots on flanks; slender, blue-gray bill; no hind toe

DIET
Insects and seeds

BREEDING
Age at first breeding: less than 1 year; breeding season: April–August (Europe and North Africa), all year (Asia and sub-Saharan Africa); number of eggs: usually 4; incubation period: 12–14 days; fledging period: 18–20 days; breeding interval: 1 year

LIFE SPAN
Up to 9 years in captivity

HABITAT
Open country with scrub

DISTRIBUTION
Southern Spain, northern and sub-Saharan Africa and southern Asia

STATUS
No more than 10 pairs left in Europe; locally abundant in parts of sub-Saharan Africa

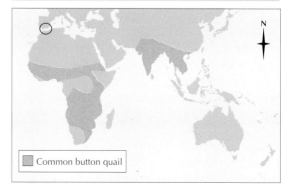

Common button quail

BUZZARD

This is a large hawk with soaring flight, distinguishable by its broad wings and rounded tail. Most of the many species of buzzards are found in open country in the Tropics. All of these species are large, with wingspans of 3–5 feet (1–1.5 m). The females are larger than the males.

The common buzzard, *Buteo buteo*, of Europe, which averages just under 2 feet (0.6 m) in length, also ranges across northern Asia to Japan, occupying a variety of habitats, from farmland to mountains and rocky coasts. Other buzzards include the rough-legged buzzard, *B. lagopus*, which breeds in northern Europe, Asia and North America, and the augur or jackal buzzard, *B. rufo-fuscus*, of Africa, which occurs at altitudes of up to 17,000 feet (5,200 m). In the United States members of the genus *Buteo* are named hawks, whereas the term buzzard is popularly used to describe vultures. The honey buzzards belong to the genus *Pernis*.

Effortless flight

Buzzards are well known for their soaring flight, in which they circle around in great spirals, with wings outstretched and the primary wing feathers separated like fingers. They circle in rising air currents called thermals up to several hundred feet high. Then they glide away until they find another thermal in which to soar. In this way buzzards can travel many miles without expending energy on flapping their wings.

It is common for birds of prey on migration to use this form of travel, especially when crossing a narrow strip of water such as the Strait of Gibraltar. They climb as high as possible over the coast, then glide across the water where there are no thermals, gradually losing height until they reach land on the far side. In Britain common buzzards do not migrate, although young birds sometimes travel across the country after leaving their parents. Scandinavian common buzzards migrate south across the Baltic Sea for the winter.

Common buzzards breed over most of Europe, except for the far north of Scandinavia. In Britain the birds breed mainly in the western half of the country, from Cornwall in the south-west to northwestern Scotland, where they are a familiar sight. In Northern Ireland common buzzards increased from just 43 pairs in 1978 to over 150 pairs by 1991. However, in the 19th century common buzzards in Britain suffered severely from the guns of gamekeepers, and by the end of that century they could no longer be seen in many areas. The advent of World War I meant that there were far fewer gamekeepers in the

The common buzzard often spots prey by waiting on a prominent perch and scanning the ground for movement.

Common buzzards fiercely defend their kills from one another.

COMMON BUZZARD

CLASS	**Aves**
ORDER	**Falconiformes**
FAMILY	**Accipitridae**
GENUS AND SPECIES	***Buteo buteo***

ALTERNATIVE NAMES
**Eurasian buzzard; steppe buzzard
(*B. b. vulpinus* only)**

WEIGHT
**Male: 1¼–2¼ lb. (0.6–1 kg);
female: 1¾–2¾ lb. (0.8–1.3 kg)**

LENGTH
**Head to tail: 20–22 in. (51–57 cm);
wingspan: 44–50 in. (113–128 cm)**

DISTINCTIVE FEATURES
**Medium-sized, thickset body; fairly broad
wings; rounded head; short tail; brown
above and paler below, but plumage varies
widely according to region**

DIET
**Mainly rabbits and other small mammals;
also birds, reptiles, amphibians and insects**

BREEDING
**Age at first breeding: 3 years; number of
eggs: usually 2 to 4; incubation period:
33–35 days; fledging period: 50–55 days;
breeding interval: usually 1 year**

LIFE SPAN
Not known

HABITAT
**Woodland edge, farmland and hills up to
about 330 ft. (1,000 m)**

DISTRIBUTION
**Britain east through Europe and Central
Asia to Japan and south as far as
northwestern Africa**

STATUS
**Common in most parts of range;
population: about 900,000 in Europe**

Common buzzard

country, which allowed the number of common buzzards to rise. However, in 1954, rabbits, an important food for common buzzards, were struck with the disease myxomatosis, and many animals that had preyed largely on them went hungry. Since then the numbers of common buzzards in Britain have again declined, though it is now thought that the numbers are fairly steady.

Patient hunters

Buzzards hunt by pouncing on animals on the ground. They sit motionless for hours on a post, tree or rocky crag, keenly watching the ground. On seeing something move, they launch themselves into a low, descending flight on half-closed wings to seize their prey on the ground by a sudden pounce. They also hunt on the wing, sometimes hovering with the persistence of a kestrel. In addition to rabbits, buzzards eat voles, mice, shrews and even moles, which are caught when they venture above ground. Frogs, toads, lizards, snakes and insects are also eaten, as are larger animals such as hares.

Many birds have been recorded as being caught by buzzards, from pheasants, crows and little owls to buntings, blackbirds and skylarks. However, buzzards can only catch other birds when the victims are taken by surprise, as they are not fast fliers. Carrion is also eaten at times. The Australian black-breasted buzzard, *Hamirostra melanosternon*, which is also known as the black-breasted kite, has the habit of driving ground-nesting birds such as emus off their nests in order to eat their eggs. The eggshells of the emu, however, are too strong for the bird to crack with its bill alone. Instead the black-breasted buzzard flies up into the air with a stone in its claws which it drops on the emu's nest. The bird then descends to eat the smashed eggs. It is unusual for an animal other than a primate to use tools, that is, to use objects to carry out tasks that the body alone cannot perform.

There are no true buteos in Australia. The black-breasted buzzard is the country's only buteolike bird of prey.

Courtship aerobatics

In the early stages of the breeding season, birds of prey indulge in graceful aerobatics, soaring, tumbling and looping the loop, to advertise their presence to other members of the species. A frequent sight in the breeding season is a pair of common buzzards gliding in circles around each other with their wings held up at an angle to form shallow V-shapes and their tails spread. The male glides a little above the female and both face one another. This display flight shows that there is a nest nearby. At other times the display takes the form of a repeated dive. The buzzard swoops steeply down with half-closed wings, then climbs again, almost vertically, to repeat the maneuver.

Rearing the chicks

The buzzard's nest is typically a large, bowl-shaped structure of sticks, heather, bracken, moss, seaweed or other material, depending on locality, built in a tree, on a rocky ledge or even on the ground on bush-covered hillsides. It is often decorated with fresh sprigs of foliage or ferns, which are regularly replaced.

Normally two to four eggs are laid, but up to six have been recorded. Very old birds may lay only one. They are incubated for 33–35 days. At first only the male brings food, while the female stays by the nest guarding the chicks. The male leaves the food by the edge of the nest, where it is picked up by the female and fed to the young. When the chicks are about 1 week old, both parents go out foraging and sometimes bring back more food than the chicks can eat. The chicks leave the nest when 6 or 7 weeks old but are fed by the parents while they learn to hunt.

The red-tailed hawk, B. jamaicensis, is one of the most familiar American raptors. It is found from Alaska to the Caribbean in all manner of habitats, including city centers and deserts (below).

CADDIS FLY

Adult caddis flies are brown in color and fly mainly at night. They can be confused with moths, but caddis fly wings are covered with tiny hairs and those of moths are covered with scales.

CADDIS FLY IS THE COMMON name given to the insect order Trichoptera, (literally "hairy wings"). There are known to be 10,000 species of caddis fly throughout the world, and it is thought that there may be as many as 50,000 species in total. Their nearest relatives are the Lepidoptera, butterflies and moths, but the wings of caddis flies are hairy instead of scaly. The antennae are long and many-jointed. The adult insects look like moths and fly at night.

Most larvae are aquatic, living in fresh water and breathing by external gills on the sides of the abdominal segments. These are the well-known caddisworms, which build tubular cases to protect their bodies, although not all caddis fly larvae do this. All larvae, however, spin silk. A few species in the family Chathamiidae, from New Zealand and Australia, are unusual for insects in having marine larvae.

Underwater builders

By far the most interesting feature of the caddis fly is the lifestyle of the aquatic larvae, which varies in the different families and genera. They can be divided into two types, the first of which are those that build portable cases. These are almost all detritivorous, that is, they feed on detritus and any food particles that float by. The second type are those that live free and are at least partly carnivorous. The case-builders use many materials in various ways to build their tubes. A gland near the mouth produces a sticky silk thread that the larva spins around itself and to which it fixes the case-building materials. Members of the genus *Phryganea*, which includes the largest caddis flies, cut pieces of leaves and stick them together with the silk. The most familiar cases are probably those of *Limnophilus*, which are made of small stones and pieces of plant stems or empty snail shells. If removed from their cases and given beads or similar objects, some of these caddis flies will use the artificial material to make new ones. *Stenophylax* and *Heliopsyche* use fine sand grains to make their cases. The former produces a straight cylinder, the latter a spiral tube that resembles a small snail shell. Cases made of stones or sand often have their weight reduced by a bubble of air trapped inside.

The cases are tubular and are always open at both ends. At the front the larva pushes out its head and thorax to move about or feed. The rear end is covered with a silk mesh. All caddis fly larvae have a pair of hooked limbs at the back, which the larvae use to hold on to the case. They hold on so tightly that attempts to pull the larvae out invariably injure them.

Most of the larvae with nonportable cases live in silken tubes, in flowing water, some living under stones in swift upland streams. In the genus *Plectronemia* the larva is nearly 1 inch (2.5 cm) long and makes a silk tunnel with the open end facing upstream, widely flared to form a trumpet-shaped net. Any small animal or piece of plant material carried into this trap by the current is seized and eaten by the larva. A number of other stream-dwelling caddis larvae make nets of various shapes to gather food. When they are damaged or choked with inedible material, the larvae clean and repair them.

The mouthparts of adult caddis flies are vestigial (imperfectly developed) and the insects can take only liquid food. In the wild they probably feed from flowers with exposed nectaries but in captivity they will take sugar and water. Case-bearing larvae eat mainly the leaves and stems of live plants. A cabbage leaf tied to a

CADDIS FLIES

CLASS	**Insecta**
ORDER	**Richoptera**
FAMILY	**Phryganeidae, Philopotamidae, Limnephilidae, Brachycentridae, Chathamiidae and many others**
GENUS AND SPECIES	**At least 10,000 species in many genera, including *Phryganea***

ALTERNATIVE NAME
Sedge fly

LENGTH
Up to 1 in. (2.5 cm)

DISTINCTIVE FEATURES
Adult: usually dull brownish; 2 pairs of wings, held over body at rest; wings covered with tiny hairs. Larva: hardened head with biting jaws; well-developed legs; many species inhabit tubular cases.

DIET
**Adult: mainly flower nectar.
Larva: vegetable matter and carrion; some species predatory.**

BREEDING
Breeding season: spring and summer; hatching period: 2–3 weeks

LIFE SPAN
1 year

HABITAT
Adults normally found near water; young stages primarily aquatic

DISTRIBUTION
Worldwide

STATUS
Most species abundant

Most caddis fly larvae live in fresh water and build elaborate structures that afford protection or serve as food-gathering devices.

string, thrown into a pond and left for a few hours, will often be covered with case-bearing caddis larvae when carefully removed. The large case-bearing larvae of *Phryganea* catch and eat water insects as well as plant food. Most tube-dwelling or free-living larvae have a mixed diet.

Life history

Caddis fly eggs are laid by the females in spring and summer. Some species drop their eggs on the surface of the water as they are flying over it; others crawl underwater and stick them to stones or plants in a jellylike mass. Some of the larvae do not make cases or tubes until they have molted their skins several times; others make tiny cases as soon as they hatch. When a larva is fully grown, nearly a year later, it pupates. Case-bearing species pupate inside the case, whereas other species undergo this transformation inside a silken cocoon.

When the time comes for the adult insect to emerge, the pupa bites its way out of the case or cocoon, using its strong mandibles, and swims to the surface of the water. Once at the surface the pupa splits open, releasing the adult caddis fly, which can fly almost immediately on emergence. The caddis fly's life cycle usually takes a year to run its course, of which the adult life represents only a small fraction.

Natural pollution detectors

Caddis flies are of great importance in freshwater ecology, with both the adults and larvae being eaten in large numbers by fish and various water birds. Caddis fly larvae are also very useful as biological indicator organisms for assessing water quality and pollution levels. They have been extensively used for this purpose both because the larvae of different species vary in their sensitivity to various types of pollution and because the taxonomy (classification) of the group is relatively well known for temperate regions. Scientists check which caddis fly species are present in a particular stream, river or lake and then compare their results with the list of species that they would have expected to find there. If certain species of caddis flies prove to be absent or present in unusually high or low numbers, this may indicate the level of pollutants present in the water.

CAECILIAN

Caecilians are widespread in the world's tropical forests. This gray caecilian, Dermophis mexicanus, is found in Costa Rica.

THE CAECILIAN IS A limbless amphibian with a long cylindrical body which is marked with rings. It lives wholly underground. Caecilians occur in warm regions, in the Americas from Mexico to northern Argentina, in southern and Southeast Asia, in the Seychelles and in parts of Africa. The 158 species are wormlike or snakelike according to size, the smallest caecilian being only 4 inches (10 cm) long, the largest, 5 feet (1.5 m). Most caecilians are about 1 foot (30 cm) in length. They are usually blackish in color but may also be pale brown, yellow or even striped.

Caecilians are smooth and slimy but, unlike other amphibians, the skin of most species has small scales embedded in it. An inner layer of skin features numerous glands that contain mucous and poison. The latter may be toxic to predators, and even to humans. It is possible that the bright colors of some caecilians, such as the orange *Schistometopum thomense*, are intended to alert would-be predators to the creatures' poisonous glands. As in snakes, one lung is large and long, the other is reduced to a small lobe.

A caecilian's eyesight is usually poor and its underground lifestyle renders this sense relatively useless. The eyes are quite small and covered over with skin and sometimes bone.

The caecilian possesses an unusual organ of smell. A retractable tentacle on each side of the head lies in a groove running between the eyes and the tip of the snout. Presumably it is protruded when the animal is moving underground. Burrows are made in soft earth, and a caecilian seldom ventures above ground except when heavy rain floods the burrows. Several species are aquatic, and some live in leaf litter.

Feeding and breeding

Earthworms are probably the main diet for most caecilian species, though some also eat termites. The sticky caecilian, *Ichthyophis glutinosus*, of Southeast Asia, the best-known species, also eats small burrowing snakes. Caecilians are eaten by certain large burrowing snakes and some birds.

There is little external difference between male and female caecilians. Little is known about mate recognition or courtship rituals, although a wriggling pre-mating "dance" has been observed in aquatic species. Caecilian fertilization is internal. The body of the male caecilian ends in a cloaca (a chamber into which the urinary, intestinal and generative canals feed). The male extends his cloaca into a vent on the female, directly transferring sperm into her body, and fertilization takes place. Some primitive species

COMMON CAIMAN

CLASS	**Reptilia**
ORDER	**Crocodylia**
FAMILY	**Crocodilidae**
SUBFAMILY	**Alligatorinae**
GENUS AND SPECIES	***Caiman crocodilus***

ALTERNATIVE NAME
Spectacled caiman

LENGTH
Up to 8 ft. (2.5 m)

DISTINCTIVE FEATURES
Bony ridge across nose; pale-colored skin around eyes

DIET
Insects, crustaceans and mollusks when young; fish, amphibians, reptiles and birds when older. Large individuals also take large mammals, such as pacas and porcupines.

BREEDING
Age at first breeding: when 3–5 ft. (1–1.5 m) long; breeding season: mostly at end of dry season, August–November; number of eggs: 15 to 40; hatching period: up to 100 days; breeding interval: 1 year

LIFE SPAN
Not known

HABITAT
All lowland aquatic habitats in range, such as rivers, ponds, swamps and creeks

DISTRIBUTION
Southern Mexico south to Peru, Bolivia and Brazil, especially in Amazon Basin

STATUS
Abundant, despite some international trade in skins; population: probably tens of thousands

Common caiman

A common caiman basking at the water's edge. Caimans raise their body temperature by sunbathing in the early morning and late afternoon, but retreat to cool water during the heat of the day.

Attacks on humans are very rare, though caimans are said to kill many domestic animals. Caimans themselves have been aggressively hunted. In some parts of their range, caimans congregate in small pools during the dry season and are killed in great numbers by cowboys.

Prey killed by drowning

Prey is sometimes seized on land and carried back to water to be held under until drowned. One technique used for catching a water bird or mammal coming down to drink is to turn and move away, submerge and then swim around to capture the prey. When swallowing, a caiman stretches its neck out of water, maneuvering its victim to a head-foremost position in the mouth. The hind foot is used to tear the food.

As with most crocodilians, caimans feed on freshwater crustaceans when small, graduating to amphibians, fish, reptiles, birds and mammals as they mature. The common caiman will even prey on piranhas. An exception to this general rule is the Paraguayan caiman, which seems to live largely on giant water snails.

Sustainable harvesting

Caimans appear to have few natural predators. The jaguar is said to kill young or half-grown caimans, and they have been found in the stomachs of anacondas. They are often killed illegally by humans and large numbers of caiman skins are still traded internationally, despite the fact that some countries have declared it illegal to hunt certain species. The black caiman, for

A black caiman hatchling on the Mamiraua Sustainable Development Reserve in Brazil. Hunted for its skin, the black caiman's population fell by 99 percent in the 20th century.

instance, is officially protected in French Guiana. However, when sold for food it is not unusual for caimans' heads to be removed, and in this way black caimans may be passed off as the near-identical common caiman, which may be legally traded.

In the face of illegal poaching, strictly monitored harvesting programs have allowed caiman populations to increase in many areas and have ensured the long-term survival of these species and their habitat. Sustainable harvesting is probably the most successful tool presently used by conservationists to maintain species, their habitats and, in some cases, entire ecosystems.

Mud nests

After mating the female caiman builds a nest of vegetation and mud scraped together in a mound near water, which she consolidates by crawling over it. Then she digs a hole in the top of the mound, where she lays between 10 and 60 eggs, depending on the species.

The eggs vary from the size of a hen's egg to the size of a goose's egg, again depending on the species. Caiman eggs are hard-shelled and the baby caiman must break the shell with the egg tooth on its snout in order to emerge. After it has been used, the egg tooth is shed. Just before hatching, the baby caimans start to croak.

On hearing this, the mother, which has remained near the nest, begins to scrape the top from the nest to assist in their escape.

The young caimans are similar in appearance to the parents, though they have relatively larger eyes and shorter snouts. In most species they are colored like the adults: dark or olive brown on the back, lighter on the flanks and dull white on the belly, with various dark blotches and patches. The black caiman is glossy black when adult, with a white or yellow underside, but the young are black with yellow bands.

Baby caimans are approximately 9 inches (23 cm) long at hatching but may grow rapidly to 2 feet (60 cm) by the end of the first year, possibly reaching 5 feet (1.5 m) at the age of 5 years.

Hunt mainly by taste

Caimans appear to be able to recognize colors, can distinguish the outlines of large objects up to 33 feet (10 m) distant and can detect sharp movements made 100 feet (30 m) or more away. The main sense organs used in hunting are the taste buds on the tongue, used when in the water, and sight, used when the caiman is above or out of water. Caimans close both their ears and their nostrils when submerged. Their eyes are adapted to night vision, having a slit pupil by day and a wide-open rounded pupil at night.

CAMEL

THERE ARE TWO SPECIES of camels: the Arabian, *Camelus dromedarius*, which has one hump, and the Bactrian, or two-humped, *C. bactrianus*. The first is not known as a wild animal, although the second survives in the wild in the Gobi Desert in Mongolia and the Taklimakan Desert in western China.

The dromedary is a special breed of the one-humped camel, used for riding, although the name is commonly and erroneously used to denote that species as a whole. The Bactrian camel is so called because it was once thought to have originated in the region of Bactria, to the north of Afghanistan.

Camels have long legs and a long neck, coarse hair and tufted tails. Their feet have two toes united by a tough web, with nails and tough, padded soles. Dromedaries develop leathery callosities on their knees through kneeling down for loading. The combined length of the head, neck and body may reach 11 feet (3.5 m); the tail is usually 13–22 inches (35–55 cm)

long, the shoulder height is up to 7½ feet (2.3 m) and the animals weigh up to 1,520 pounds (690 kg). Bactrians are swift, with smaller feet and no knee pads. Their coat is brownish red and short, and their ears are smaller than those of domesticated camels. Bactrians' humps are smaller than the dromedaries' single hump.

The camels' unusual dentition enables them to give a powerful bite. At birth they have six incisors in both upper and lower jaws, a canine on each side, then a premolar followed by a gap before the cheek teeth. As camels grow, they lose all except the outside incisors of the six in the upper jaw; these gradually take on a shape similar to the canines.

There are several theories about the derivation of the phrase "ship of the desert," a term often used to describe the camels. One suggestion relates to the animals' characteristic gait. Like the giraffe, a camel moves both the foreleg and the hind leg on one side of its body when it walks. This action gives the camel a distinctive

Camels are well adapted to the high temperatures of deserts and can survive for several weeks without drinking water, provided they have access to succulent plants.

swaying motion, reminiscent of a ship at sea. Reports suggest that camels are also capable of swimming, although the infrequency of such sightings probably means that the animals are reluctant to do so.

Habits

The wild Bactrian camels of the Gobi Desert are active by day, associating in groups that may comprise 30 or more individuals. Except for one male, these groups are exclusively female.

Young camels closely resemble adults, apart from their incisors, their soft fleeces and lack of humps. The young Arabian camel also lacks knee pads. A single calf, or on rare occasions two, is born 370–440 days after conception. Its only call is a soft *baa*. It can walk freely at the end of the first day but is not fully independent until 4 years old and becomes sexually mature at about 5 years. The maximum recorded life for camels is 50 years.

Adaptations to desert life

A camel's external features and its physiology are specifically adapted to life in deserts. Its eyes have long lashes, which protect them from windblown sand. The camel's nostrils are muscular and can be readily closed, or partly closed, to keep out sand. The long neck and legs provide a large surface area relative to the volume of the animal's body. This facilitates heat loss, enabling the camel to remain cool in high desert temperatures.

The camels' physiology features other adaptations that provide protection from overheating and help the animals to withstand dehydration and physical exertion with a minimum of feeding and drinking. It has been shown that camels on a completely dry diet can go several weeks without drinking, although they lose water steadily through their skin and their breath as well as in their urine and feces. However, most desert journeys are made in winter when more water and vegetation containing water is available. Normally, a camel feeds on desert plants with a high water content.

Do camels store water?

Camels are able to drink 27 gallons (100 l) of water, or more, in 10 minutes, although they will do so only to replenish the body supply after intense dehydration. In those 10 minutes a camel will pass from an emaciated animal, showing its ribs, to a normal appearance. Few other animals can effect such a rapid transformation in their body shape. The water does not stay in the stomach: it passes into the tissues, and a camel appears swollen after a long drink. For many centuries it was thought that camels stored water

CAMELS

CLASS	**Mammalia**
ORDER	**Artiodactyla**
FAMILY	**Camelidae**

GENUS AND SPECIES **Arabian camel, *Camelus dromedarius*, Bactrian camel, *C. bactrianus***

ALTERNATIVE NAMES
C. dromedarius: one-humped camel; dromedary. C. bactrianus: two-humped camel.

LENGTH
Head and body: 7–11 ft. (2.25–3.5 m); tail: 13–22 in. (33–55 cm)

DISTINCTIVE FEATURES
Long neck, legs and eyelashes; short, coarse hair; tough, padded soles to feet; hump on back (1 in *C. dromedarius*; 2 in *C. bactrianus*)

DIET
Desert vegetation; rarely carrion

BREEDING
Age at first breeding: 5 years; breeding season: variable; number of young: usually 1; gestation period: 370–440 days; breeding interval: about 2 years

LIFE SPAN
Up to 50 years

HABITAT
Arid land

DISTRIBUTION
C. dromedarius: throughout North Africa and Middle East; feral populations in Australia. C. bactrianus: only wild populations in Great Gobi National Park, Mongolia, and Taklimakan Desert, China.

STATUS
C. dromedarius: common in domesticated and feral state. C. bactrianus: endangered; 500 to 600 in Great Gobi National Park, 380 to 500 in Taklimakan Desert.

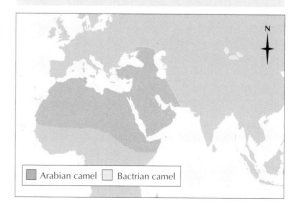

Arabian camel   Bactrian camel

in their stomachs. Scientists now know that pockets in the camel's stomach lining contain liquefied masticated food rather than water.

A camel can lose water equal to 25 percent of its body weight and show no signs of distress. In contrast, a human losing 12 percent of body water is in great distress because this water is drawn from tissues and blood. The blood becomes thick and viscous, and the heart has greater difficulty in pumping it through the veins. A camel loses water from its tissues but not from the blood, so there is no strain on the heart. Thus, an emaciated camel is capable of the same physical exertion as it would if under normal circumstances. The mechanism for this is not known. The only clear difference between the blood of a camel and that of other mammals is that its red corpuscles are oval, not discoid.

The camel's hump

The hump contains a store of fat. It has been argued that this can be converted to water, and that the hump effectively acts as a water reserve. The average hump of the Arabian camel may

contain as much as 100 pounds (45 kg) of fat and 15 gallons (57 l) of water. To convert this fat to water, however, extra oxygen is needed, and scientists have calculated that the breathing needed to get this extra oxygen would itself lead to the loss of more than 15 gallons of water as vapor in the camel's breath. It follows that the fat stored in the hump is broken down to supply energy rather than water; the water released as a by-product of this process is lost. The camel's hump thus effectively acts as a reserve of energy.

During the summer a camel excretes less urine and, more importantly, perspires very little. The camel's highest body temperature is 105° F (40° C) by day but during the night this drops to 93° F (34° C). Human body temperature remains constant at approximately 98½° F (37° C) and as soon as the day starts to warm up the heat is felt. A camel starts with a temperature of 93° F at dawn and does not heat up to 105° F until nearly midday. The camel's coarse-haired coat provides insulation against the heat of the day and keeps the animal warm during the bitterly cold desert nights.

Camels will feed on any desert vegetation they can find. These Bactrian camels are browsing in desert outside Bukhara, Uzbekistan.

Camels have long eyelashes to prevent windblown sand from entering their eyes. Likewise, their muscular nostrils can be fully or partly closed to prevent the inhalation of sand particles.

Origins of the camels

Camels originated in North America, where many fossils have been found, small and large, with short necks or long, as in the giraffelike camels. The smallest was the size of a hare; the largest stood 15 feet (4.5 m) at the shoulder. As the species multiplied there was one migration southward into South America and another northwestward, and then across the land-bridge where the Bering Strait is today, into Asia. As the numerous species died out over the last 45 million years, the survivors remained as the South American llamas and Asian camels.

A few species reached eastern Europe and died out. None reached Africa. Until 6,000 years or more ago there was only the one species in Asia, the two-humped Bactrian camel. It is not known when the one-humped camel came into existence, but evidence suggests that it is a domesticated form derived from the Bactrian camel. Both species readily interbreed, and the offspring usually have two humps, the hind hump being smaller than that in front.

The first record of a one-humped camel is on pottery from the sixth dynasty of Ancient Egypt (about 3500 B.C.E.). This is surprising because the camel was not known in the Nile Valley until 3,000 years later. Its representation on the pottery may have been inspired by a wandering camel train from Asia Minor. The camel appears on Assyrian monuments dated about 1000 B.C.E., and from then onward. When the Queen of Sheba visited King Solomon in Jerusalem in 955 B.C.E., she brought with her several one-humped camels. The name camel seems to have been derived from the Semitic word *gamal* or *hamal*, meaning "carrying a burden."

Feral camels

Experts believe that the one-humped camel was selectively bred from domesticated two-humped camels in central Asia by peoples who left no records. Today the Bactrian camel is confined to Asia but most of the 3 million Arabian camels are on African soil. Some, however, have been introduced into countries far from Africa or Asia. In 1622 some camels were taken to Tuscany, Italy, where a herd still lives on the sandy plains near Pisa. On the plains of Guadalquivir in southern Spain there are feral camels which were taken there by the Moors earlier still.

Spanish conquistadors brought camels to South America in the 16th century, though these have subsequently died out. Other camels were taken to Virginia in 1701, and there was a second importation to the United States in 1856. The survivors of this group were still running wild in the deserts of Arizona and Nevada in 1915. Camels were taken to northern Australia, and there also they have reverted to the wild.

CANADA GOOSE

THE CANADA GOOSE IS a large, gray-brown goose with a black head and neck and a white patch extending from the chin, up the cheeks to behind the eyes. The tail is black while the feathers around the base of the tail are white. The Canada goose is sometimes confused with the barnacle goose, *Branta leucopsis*, but the latter is smaller and has a wholly white face, black breast and light underparts.

Canada geese live in Alaska, Canada and the northern parts of the United States, migrating southward as far as the Gulf of Mexico in winter. European settlers became aware of Canada geese due to the species' large gatherings in fall and spring around Newfoundland and Hudson Bay. Live birds were brought back to Europe and domesticated in the 17th century. By the 18th century they were a familiar sight in England, and bred on lakes in many country estates. Canada geese have since spread to other parts of northern Europe, and have also been introduced

to New Zealand. Unlike the barnacle goose, the Canada goose favors inland areas such as lakes, marshlands and parkland, although it is also found on estuaries and seashores. It is gregarious outside the breeding season, forming groups of 200 to 300.

Grazers on land and water

The Canada goose feeds mainly on plants. It grazes on grasslands, on the rushes and sedges in marshes and on sea lettuce, eelgrass and various algae around the seashore.

During their spring migration Canada geese feed on fields of sprouting grain. Grazing the shoots does no permanent damage to the crop, though sometimes the seeds are unearthed. The geese also feed on swelling buds of trees and bushes at this time. After the breeding season, when the geese are moving south, they graze in stubble fields. Canada geese have favorite feeding grounds. They generally congregate in areas

North America's most common goose, the Canada goose pairs for life and is fiercely territorial when nesting.

This day-old gosling will achieve full adult size and coloration within 6 weeks.

CANADA GOOSE

CLASS	**Aves**
ORDER	**Anseriformes**
FAMILY	**Anatidae**
GENUS AND SPECIES	***Branta canadensis***

WEIGHT
8–12 lb. (3.7–5.4 kg)

LENGTH
Head and body: 22–43 in. (56–110 cm); wingspan: 48–72 in. (122–183 cm)

DISTINCTIVE FEATURES
Large gray-brown goose with black neck and head; white saddle patch from chin to cheek

DIET
Mainly plant material; some worms, insects, snails and crustaceans; rarely small fish

BREEDING
Breeding season: March–June; number of eggs: usually 5 or 6; incubation period: 28–30 days; fledging period: 40–45 days; breeding interval: 1 year

LIFE SPAN
Up to 23 years

HABITAT
Marshes, lakes, grasslands, coastal wetlands

DISTRIBUTION
Native to Canada, U.S., southern Greenland and northeastern Russia; introduced to northern Europe and New Zealand

STATUS
Common throughout range

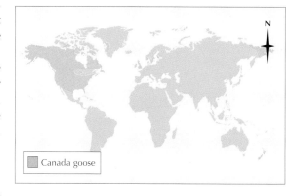

Canada goose

where food is abundant at that particular season. The feeding flocks are guarded by sentries that warn of approaching danger. These lookouts are relieved at intervals so that they may feed also.

During the breeding season, Canada geese turn to a more aquatic diet. They feed on water plants, worms, insects, snails, crustaceans and even small fish. When feeding in the water the geese dip their long necks into the water and upend like ducks.

Faithful partners

Canada geese mate for life and arrive at the breeding grounds already bonded in pairs. Nevertheless, courtship and fighting occur on the breeding grounds. The ganders (males) drive others from their chosen mates by advancing with neck horizontal and bill open, hissing and rustling their plumage. After the opposition has

been driven off, the gander returns to his mate and they perform a dance with sinuous movements of their necks.

Nesting begins in March or April. The precise date depends on latitude, occurring later in northern regions. It may be delayed by up to

2 weeks in bad weather. The nests are normally built on the ground near water but, especially in areas liable to flooding, they may be built on rocky ledges or in trees. In Britain Canada geese sometimes take over the deserted nests of birds of prey, such as common buzzards.

The nest is lined with grasses, reeds and feather down. The eggs are brooded by the goose (female) alone, and the gander stands guard throughout the 28–30-day incubation period. Ganders defend their offspring vigorously, although foxes will sometimes kill incubating Canada geese. As soon as the chicks have hatched, the parents lead them down to water, where they swim out, gander leading and goose bringing up the rear.

The chicks can dive at an early age. They are full grown at 6 weeks, weighing about 10 pounds (4.5 kg) and already possessing the characteristic adult plumage. Before they can fly, young geese have many predators, which attack them both from the air and from underwater. Canada geese, migrating up and down North America, are shot as they feed in the fields or as they land on lakes, lured down by decoys or tame geese. Arctic foxes also prey on Canada geese, especially in years when their normal prey, lemmings and other small mammals, is scarce.

Harbingers of spring

The arrival of Canada geese in the northern regions of North America is the traditional signal that spring has come. The month that they arrive is known to the Native Americans around Hudson Bay as the Goose Moon. Early European settlers depended a great deal on the geese, killing several thousand a year and preserving them in barrels of salt for winter use.

The migrations of Canada geese have been studied for many years. Their speed as they fly in V-shaped or W-shaped formations has been recorded as 60 miles per hour (100 km/h), although the overall movement up the continent is much slower than this speed suggests, taking as long as 2 months. The geese that have wintered the farthest south move first. They gather in restless flocks, honking to each other and preening their feathers in preparation for the long flight. At first they move 9 miles (14.5 km) a day, but this increases to 30 miles (50 km) as the impetus of the movement increases. The journey is also related to the warming of the atmosphere: it has been shown that the leading flocks of Canada geese keep level with the 36° F (2° C) temperature line as it advances north across central and northern North America. Countless other flocks of Canada geese follow not far behind

Canada geese form large and noisy groups outside the breeding season; flocks of 200 to 300 individuals are common.

CANARY

Wild canaries are notable for their bright yellow or greenish yellow plumage and distinctive song. Captive birds (above) are selectively bred to enhance these traits.

THE CANARY IS A finch in the subfamily Carduelinae, or typical finches. It is native to three archipelagos in the Atlantic Ocean: the Canary Islands, the Azores and Madeira. Wild canaries are mainly greenish-yellow with olive-brown streaking, stronger in the female. The first canaries to be imported into mainland Europe probably arrived in the 16th century, and selective breeding has since produced a number of color varieties, the most common being pale yellow, bright yellow, yellow tinged with orange, and mottled with brown and black. Some captive-bred varieties have a crested head.

Hybrid canaries are sometimes referred to as mules. They are readily produced with several other members of the Carduelinae, including the greenfinch, *Carduelis chloris*, goldfinch, *C. carduelis*, and siskin, *C. spinus*.

Sociable songbirds

Canaries live in bushes and clumps of trees, and regularly visit gardens, orchards and vineyards. They are sociable, with habits similar to those of other species of finch. Their popularity as cage-birds is a reflection of their distinctive song. As is

the case with many birds, the canary's song is used partly for announcing to other members of the species that a male is in occupation of a territory and partly as a signal of male fitness to listening females. The superb song is rich, varied and musical, and incorporates complex phrasing. A singing bird will chirp, pipe, trill, twitter and even wheeze, while varying both the tempo and timbre of the song. Canaries sing throughout the year, but are most active in early spring, when males perform aerial song-flights. They take off from a prominent perch and fly a short circuit, singing in midair.

Recent research into the singing behavior of the canary has produced some fascinating results. It has been shown that canaries are able to discriminate between the songs of individual birds, including those of other species as well as their own. Moreover, each female canary appears to favor a particular type of song, incorporating a particular combination of phrases and segments. This choice is influenced by the songs that the female experienced during her early life. A female will try to breed with a male that corresponds to her song preference, and will produce more sexual displays for her chosen mate's song than for the songs of other males. Researchers have also found that a special, two-note syllable in canary songs elicits high levels of sexual display in females.

Although domesticated canaries will mimic to some extent the songs of other birds and imitate human speech, most of their song is innate. A young canary hand-reared in isolation and allowed to hear only the song of a particular species of warbler will use a song intermediate between that of a canary and the warbler. If later placed with other canaries, it will tend to lose the adopted "warbler" aspects of its song.

Island birds

The canary is common within its restricted range, and occurs in most available habitats. It is resident, but tends to wander in loose flocks outside the breeding season; there is evidence of flocks moving between different islands in the Azores and Madeira archipelagos.

The canary feeds mainly on the small seeds of low-growing herbaceous plants, also taking blossoms and leaf and fruit buds. This diet is supplemented with a few insects, especially during the breeding season. Canaries frequently visit orchards and cultivated crops, especially tomatoes, and are also attracted to banana and orange trees in bud.

CANARY

CLASS **Aves**

ORDER **Passeriformes**

FAMILY **Fringillidae**

SUBFAMILY **Carduelinae**

GENUS AND SPECIES *Serinus canaria*

WEIGHT
Average ½ oz. (15 g)

LENGTH
**Head to tail: 5 in. (12.5 cm);
wingspan: 8–9 in. (20–23 cm)**

DISTINCTIVE FEATURES
**Male: mainly greenish yellow plumage, with
olive-brown streaks on head and back and
brown wings and tail. Female: duller and
more heavily streaked.**

DIET
Seeds, buds and blossom; some insects

BREEDING
**Age at first breeding: 1 year; breeding
season: eggs laid January–July, peaking
April–June; number of eggs: usually 3 or 4;
incubation period: 13–14 days; fledging
period: 15–17 days; breeding interval:
2 or 3 broods per year**

LIFE SPAN
Probably up to 8 years

HABITAT
**Open country, farmland, orchards,
vineyards and gardens**

DISTRIBUTION
**Islands of the Madeira, Azores and Canary
Islands archipelagos. Introduced to Bermuda,
Puerto Rico and Midway Islands (Hawaii).**

STATUS
**Common; population: Canary Islands,
80,000 to 90,000 pairs; Azores, 30,000 to
60,000 pairs; often escapes from captivity**

Canary

Perhaps strangely for such a well-known bird, the canary is relatively little studied in the wild. It seems to breed in neighborhood groups, with pairs defending small territories. Each pair builds a nest high up in thick shrubs or trees. The female chooses the nesting site and does most of the building. The saucer-shaped nest is made of grasses, roots and moss, lined with hair and feathers. Building activity reaches its peak 4 days before the first egg is laid and the three or four pale blue eggs are laid at 24-hour intervals. The eggs hatch in 14 days, the nestlings being at first blind and naked except for a small amount of down. The eyes open about 7 days after hatching. Feathers then begin growing, and the plumage is complete within a month. The young leave the nest in 15–17 days and become independent 36 days after hatching.

History of domestication

The group of islands now known as the Canary Islands were conquered by Spain at the end of the 15th century. It was doubtless this final conquest, after nearly a century of unsuccessful military expeditions, that led to the importation of the canary to Europe.

The precise date of the introduction of the canary into Europe is not known, although the site of the introduction is believed to have been at a place on the mainland of Italy, opposite the island of Elba. The date of introduction can only be guessed from the fact that Conrad Gesner, the Swiss naturalist, writing in 1585, mentions the canary as a cagebird, and Ulysses Aldrovandus, the Italian naturalist, writing in 1610, gives a full description of the species. The first use of the term canary-bird in England was in 1576 and the first reference to the Canary Islands was later, in 1592. Presumably the fashion for keeping canaries spread rapidly and canaries in cages became a familiar sight, because by 1673 the term canary-bird, meaning a jailbird in criminal slang, had entered the English language.

This female canary has been trapped by ornithologists as part of a research program. It will be weighed and measured before being released unharmed.

CAPE BUFFALO

Although buffalo herds were severely affected by overhunting and an outbreak of rinderpest disease in the 1890s, their numbers have since stabilized.

THE CAPE BUFFALO is the only species of wild cattle found in Africa. A powerful and heavy creature, it has the reputation of being the most dangerous of all the African big-game animals.

This bulky oxlike creature stands 3⅓–5⅔ feet (1–1.7 m) at the shoulder and adult bulls weigh up to a ton (900 kg). The head and shoulders are heavily built and support characteristic large horns that spring from broad bases, sometimes meeting at the midline of the forehead, and curving first down, then up, to finish in a point. The horns may span as much as 56 inches (1.4 m) and in some cases point back rather than up. The coat is brownish black, thick in young buffalo and sparse in older individuals. Adult bulls have a dark brown to black coat, cows have a lighter brown coloring and calves are reddish brown.

Cape buffalo range from Lake Chad south to the Cape of Good Hope and from Senegal east to the Horn of Africa. A subspecies of Cape buffalo is found in the equatorial forests from Senegal to eastern Congo. This is the forest buffalo, *Syncerus caffer nanus*, or bushcow. Smaller than the Cape buffalo, it is reddish brown in color and its much smaller horns point straight back rather than up. Its small size and backward-pointing horns are an adaptation to living in thick forests.

Buffalo have poor sight, though their sense of taste is well-developed. They are quiet animals, communicating through subdued lowing, especially in dark forests, which enables even blind buffalo to keep in contact with the rest of their herd. An exception to this behavior occurs during the mating season when buffalo will grunt and bellow to attract attention.

Depleted herds

The Cape, or African, buffalo was once common throughout Africa south of the Sahara. However, its range is more restricted today. Hunting and loss of habitat has significantly reduced numbers, and toward the end of the last century Cape buffalo contracted a disease called rinderpest

CAPE BUFFALO

CLASS	**Mammalia**
ORDER	**Artiodactyla**
FAMILY	**Bovidae**
GENUS AND SPECIES	***Syncerus caffer***

ALTERNATIVE NAMES
African buffalo (both subspecies). Forest buffalo; bushcow (forest subspecies only).

WEIGHT
⅓–1 ton (300–900 kg)

LENGTH
Head and body: 6⅔–11 ft. (2–3.5 m); shoulder height: 3⅓–5⅔ ft. (1–1.7 m). Grassland subspecies larger than forest subspecies.

DISTINCTIVE FEATURES
Grassland subspecies: broad horns, joined and pointing upward. Forest subspecies: smaller horns, not joined, pointing backward.

DIET
Mainly grasses and other ground vegetation

BREEDING
Age at first breeding: 3–5 years; breeding season: varies according to climate; number of young: 1; gestation period: about 340 days; breeding interval: usually 2 years

LIFE SPAN
Up to 18 years

HABITAT
Grassland subspecies: grassland in areas with good water supply. Forest subspecies: rain forest, including in uplands.

DISTRIBUTION
Grassland subspecies: southern and eastern Africa. Forest subspecies: rain forests of central Africa.

STATUS
Low risk; population reduced in places due to diseases and competition with domestic cattle

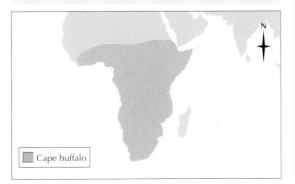

Cape buffalo

from domestic cattle. This decimated the population over the greater part of southern and eastern Africa, and in places only isolated populations which were not reached by the disease have survived. However, despite the reduction in the numbers of western and central African Cape buffalo herds, the species is not endangered. Large numbers inhabit African national parks and reserves.

Forest buffalo form small herds, usually of up to 12 animals. The core of the group is composed of related females and their offspring; one or two bulls complete the herd. Other bulls are solitary, or form smaller "bachelor" herds. Cape buffalo are often more gregarious than forest buffalo, and live in herds ranging from a few dozen to several hundred. Like forest buffalo, they live in family-related groups which are regularly extended into larger, clanlike gatherings. The clans are hierarchical, and individuals of both sexes assume different positions of rank. Although numbers are much reduced, herds 1,000 to 2,000 strong may still be seen in some places, such as the Kruger National Park in South Africa. However, gatherings of this size are only possible on large areas of rich pastureland or during rainy seasons when suitable food is plentiful.

Like many animals of tropical or subtropical regions, Cape buffalo are most active in the early morning and evening. During the heat of the day they lie in the shade, and at night they rest and chew the cud. They are never found far from water, usually restricting their grazing to within about 12½ miles (20 km) of a water source, and go down to drink at the water's edge during the morning and evening. Their need to drink all day limits the buffalo's range to areas that contain a water source.

The Cape buffalo's large horns make it one of the most dangerous big-game animals. The horns of the forest buffalo point backward and tend to be smaller and less formidable.

Most buffalo herds consist of females and their offspring, with one or two bulls. Bulls sometimes form separate all-male herds.

Dangerous when panicked

Under normal circumstances, Cape buffalo are not aggressive, although they are dangerous if panicked or when they are wounded. If the buffalo are approached cautiously across open country, so that they can see what is coming, their reaction is minimal: some of the bulls may leave the herd to investigate, lifting their muzzles and flaring their nostrils in a characteristic fashion. If it is approached unexpectedly from behind, however, a Cape buffalo herd is likely to retreat along its tracks in a panic.

The Cape buffalo is a grazer and its muzzle is flattened to square off with the ground. Herds seek out woodland, open country and valley bottoms, avoiding the heat of the day by lying in wallows, taking cover under dense vegetation or simply remaining mobile to keep cool. Forest buffalo browse leaves, twigs and young shoots as well as feeding on grass. Both Cape and forest buffalo quickly adapt their feeding patterns to suit potentially dangerous circumstances, such as human disturbances. Herds can rapidly switch from grazing all day to grazing only at dawn and dusk or during the night to avoid leaving themselves vulnerable to attack.

The time of mating varies within the buffalo's range, and in some places it occurs nearly all year round. Older bulls are often driven from the herd by younger and more powerful bulls and violent confrontation between bulls is common during the rutting season.

The single calves are born after a gestation of about 11 months, and a gap of 2 years between births is common. Most calves are born during rainy seasons. Within 10 minutes of birth the calf is able to stand up and within a few hours it is able to keep up with the herd's movements. Although most wild buffalo live for about 16 years, there are reports of some living for up to a decade longer.

Predators

Lions will prey on buffalo, although the latter are exceptionally powerful and often manage to kill their attackers. In Zambia, Cape buffalo are often the main prey of lions; usually it is the old, infirm or immature animals that are killed. Leopards will attack newborn buffalo calves, but do not attempt to confront animals that are more mature. Buffalo crossing rivers are also vulnerable to attack from crocodiles.

CAPE HUNTING DOG

THE CAPE HUNTING dog is a ferocious carnivore only distantly related to the domestic dog. Unlike true dogs, foxes, wolves and jackals, the Cape hunting dog, or African wild dog, has only four toes on the front foot, that is, the dew claws are missing. It is the largest member of the dog family to occur in Africa, standing just over 2 feet (0.6 m) high at the shoulder and measuring 4 feet (1.2 m) from nose to bushy tail.

The mottled black, white, yellow and yellowish brown coat and large, round ears give the Cape hunting dog a patchwork look; no two dogs have the same pattern. The tip of the dog's tail is usually white, and its muzzle black; a distinctive black line runs from the tip of its muzzle to between its ears. The Cape hunting dog has a strong odor and utters a distinctive call somewhat similar to the sound of an oboe or a horse's whinny.

Hunters with a base camp

The range of the hunting dog extends from south of the Sahara Desert south to Botswana, wherever there is open savanna country. The species lives in packs of 4 to 60, although usually groups consist of about a dozen individuals. The packs are nomadic, staying in one area as long as there is an abundant supply of food, then moving on. When a pack arrives in an area, it sets up a base camp, from which hunting parties set out. The "camp" is often set in the abandoned burrows of other animals or by a water hole. Here the pups are left with guardians while the rest of the pack ranges across the countryside. If food is plentiful the pack will not travel far, but as it becomes scarce the dogs hunt over a wide area. A pack's home range may be as large as 80–800 square miles (205–2,070 sq km). Packs of hunting dogs average about six adult males and four adult females today, but in the past packs included hundreds of dogs.

Teamwork

Hunting dogs work systematically as a pack, running down their prey and eating it even before it is dead. One pack that was observed for a long period in the Serengeti Plain had two regular hunting times: 6.30–7.30 (sunrise) and 18.00–19.00 (sunset).

When the dogs are at rest, their prey—mainly antelope, although small mammals such as cane rats and birds will be eaten in times of shortage—take little notice of them. The prey species will wander, grazing, to within about 300 feet (90 m) of the hunting dogs. However, as soon as the dogs begin to move, the antelope or other large mammals sense danger and flee. Gazelles, such as Thomson's (*Gazella rufifrons*) and Grant's (*G. granti*) gazelles, take flight when they see a hunting pack ½ mile (0.8 km) away, but the dogs catch up despite this head start.

No two Cape hunting dogs share exactly the same coat coloring and pattern, although there are family and regional resemblances.

Unlike the members of the cat family, which are sprinters, giving up if the quarry is not caught after the initial burst of speed, dogs are long-distance runners. Hunting dogs are able to chase prey over distances of up to 3 miles (5 km) at a steady 30 miles per hour (48 km/h); they can exceed 37 miles per hour (60 km/h) for short spurts of just over 1 mile (1.6 km). Cape hunting dogs have great staying power and usually run down their prey after a long chase. Frequently the prey jinks to avoid its immediate pursuer but this will bring it closer to another dog, which will be able to slash at it with its canine teeth. The quarry is worn down by the dogs snapping at its flanks. As a result, it soon becomes weak from loss of blood and the pack descends on the victim. First to be eaten are the entrails, which are rich in vitamins and other essentials, but eventually there is little left except the skull.

Wildebeest do not run from hunting dog packs. The bulls move forward and try to repel the dogs by charging them, but the dogs usually avoid these attacks and dodge around the bulls to where the calves and cows are huddled. Providing the wildebeest stay in a tight bunch, they are safe from the hunters, but as soon as a calf becomes separated it is seized.

Cape hunting dogs are limited in some respects by food. Their hunting methods use up a great deal of energy, and failed hunts put the hunting party at a great disadvantage. Kleptoparasitism (food stealing) by spotted hyenas, *Crocuta crocuta*, raises the hunting costs

Cape hunting dogs prey mainly on antelope but will take smaller creatures, including birds and rats, if given no alternative.

CAPE HUNTING DOG

CLASS	**Mammalia**
ORDER	**Carnivora**
FAMILY	**Canidae**
GENUS AND SPECIES	***Lycaon pictus***

ALTERNATIVE NAMES
African wild dog; African hunting dog

WEIGHT
37½–79 lb. (17–36 kg)

LENGTH
Head and body: 30–44 in. (75–112 cm); shoulder height: 24–31 in. (61–78 cm); tail: 12–16 in. (30–41 cm)

DISTINCTIVE FEATURES
Multicolored coat with blotchy pattern unique to each individual; rounded ears; bushy tail with white tip; 4 toes on each foot rather than 5 as in true dogs

DIET
Mainly large mammals such as antelope; some rodents and birds

BREEDING
Age at first breeding: 21–60 months (male); at least 22 months (female); breeding season: all year, with peak in December; gestation period: about 79–80 days; number of young: usually 6 to 8; breeding interval: usually 11–14 months

LIFE SPAN
Up to 10 years

HABITAT
Grassland, savanna and open woodland

DISTRIBUTION
Much of sub-Saharan Africa

STATUS
Critically endangered in many countries; vulnerable elsewhere. Population: probably less than 5,000.

Cape hunting dog

significantly and can indirectly lead to high mortality or low productivity in packs. The dogs usually hunt for about 3½ hours a day, but must increase this to 12 hours if they lose 25 percent of their food to the stronger hyenas.

Babysitting duties

The closely knit nature of hunting dog packs is reflected in their breeding behavior. Pups are born blind and at any time of the year after 79–80 days' gestation. Because of the dogs' communal breeding habits it has been difficult to establish how many pups there are in a litter. Observations at zoos suggest that there may be as many as 19, although six to eight would be nearer the average number. The pups are brought up in a communal burrow, an abandoned aardvark hole for instance, and the adults share the duty of rearing them.

When the pack goes hunting, some females and certain males stay behind to guard the pups. On returning, the hunters disgorge some of their food to the pups and the guards. When the pups are very small, the females beg for extra meat from the hunting party and chew it up to give to the pups at intervals. The pups emerge from

their burrow after 1 month and are weaned off their mother's milk to begin eating regurgitated meat about a week later. In common with other young carnivores, the pups spend much time in play simulating their hunting behavior and will start to accompany adults on hunts when they are about 3 months old. In a pack, the hunting dogs need fear nothing except humans, but lions sometimes investigate nursery burrows, which results in the pups being moved away.

An endangered species

Cape hunting dogs have undergone possibly the most dramatic range contraction of any African carnivore. Due to a comparatively short life span of about 10 years, the species depends on the speedy renewal of packs and unimpeded access to plentiful supplies of prey in order to ensure its survival. However, the destruction of its natural habitat to provide grazing for livestock, the spread of diseases from the growing populations of domestic dogs and persecution from humans has resulted in a marked decrease in the numbers of Cape hunting dogs. They are classified as endangered in many countries and are under increasing threat in the rest of their range.

Blind and helpless at birth, hunting dog pups are quickly assimilated into their packs and accompany adults on hunts after only a few months.

CAPERCAILLIE

During its courtship display the male capercaillie lifts up its head, revealing its distinctive beard, and fans out its tail feathers. These actions are accompanied by strutting movements and a variety of calls.

THE CAPERCAILLIE IS the largest member of the grouse family, the males being nearly 3 feet (90 cm) long and the females 2 feet (60 cm). Its name is also spelled "capercailzie," a word derived from the Gaelic for bird or old man of the wood.

The general color of the male capercaillie's plumage is a dark slate gray with flecks of white on the body. The throat and sides of the head are black and the breast is glossy greenish black. There is some bright red skin over the eye, the bill is whitish and there is a beard of greenish black feathers, similar to an inverted crest, under the chin. The female is browner and is sometimes confused with one of the smaller species of grouse, though it has a reddish orange patch on the breast. Capercaillie sometimes interbreed with black grouse, *T. tetrix*, but usually the two species live in different localities and do not mix.

Found in coniferous forests

The range of the capercaillie has diminished over time, since coniferous forests, its traditional habitat, have been cut down, although in some places the species has spread into new plantations.

Its present range includes Scandinavia and eastern Europe across to Lake Baikal, roughly 52–66° N. This area corresponds approximately to the July temperature lines of 53° F (12° C) in the north and 70° F (21° C) in the south. Capercaillie are also found in parts of central Europe and the Balkans, Austria, Switzerland, southern Germany and eastern France. There are isolated populations in the Pyrenees and central and northeastern Scotland. A second species, the black-billed capercaillie, *T. parvirostris*, is a rare bird restricted to northeastern Asia.

The original populations in Ireland and Scotland were exterminated in the 18th century when their forest habitats were destroyed. In 1837 and 1838 some capercaillie were imported from Sweden and released at Taymouth in Scotland. However, capercaillie have been declining again in the western parts of their range, including Scotland, since the 1970s. Efforts are now being made to reintroduce populations into commercial conifer plantations in Scotland.

The flight of the capercaillie is that of a typical gamebird: rapid wingbeats followed by a glide on down-curved wings. When flushed,

CAPERCAILLIE

CLASS	**Aves**
ORDER	**Galliformes**
FAMILY	**Tetraonidae**

GENUS AND SPECIES **Western capercaillie,** ***Tetrao urogallus*; black-billed capercaillie,** *T. parvirostris*

ALTERNATIVE NAMES
Capercailzie; horse of the woods (archaic)

WEIGHT
3½–5½ lb. (1.5–2.5 kg)

LENGTH
Head to tail: 24–34 in. (60–87 cm)

DISTINCTIVE FEATURES
Largest grouse in region. Male: slate gray or black with patch of red skin over eye; whitish bill; distinct beard; glossy, greenish black breast. Female: smaller and browner, with reddish orange breast.

DIET
Berries, buds, flowers, seeds, pine needles, pine cones, twigs and grasses; some insects

BREEDING
Breeding season: April–June; number of eggs: usually 7 to 11; incubation period: 24–26 days; fledging period: 14–15 days; breeding interval: 1 year

LIFE SPAN
Up to 10 years

HABITAT
Boreal and temperate forests, particularly coniferous

DISTRIBUTION
Parts of western, central and eastern Europe; Pyrenees Mountains; Scandinavia and Russia east to Siberia

STATUS
Uncommon and declining in much of range; estimated population: more than 700,000

Capercaillie

capercaillie usually fly only a short way before landing. The takeoff is very noisy but as the flight levels off, it becomes rapid and silent, in the same manner as that of the wood pigeon, *Columba palumbo*. Capercaillie are also able to move by running rapidly across the ground.

In the summer capercaillie are generally seen singly, but in winter the males in particular become gregarious, forming large parties.

Seasonal diets

The capercaillie subsists almost entirely on plant food. A few beetles and other insects, mainly grubs, are also taken, but these are not thought to be an important part of the bird's diet.

During the summer, a wide variety of plants is eaten, such as grasses, heather flowers and berries, along with the fruits of rowan, bramble, cranberry and hawthorn and the seeds of violet and buttercup. In the fall, about mid-November, both diet and habitat change as capercaillie move from forests with little or no pine to those with

The female capercaillie is often confused with other grouse. However, it is larger and bears a distinctive reddish patch on the breast.

The male capercaillie is similar in size to a turkey and is known for its shows of aggression. It will attack dogs, sheep, deer and even humans if it feels that its territory is endangered.

pine predominating. At the same time they cease feeding on the ground and are mostly found in the trees. Their diet then consists mainly of the needles of Scots pine, together with twigs, buds and unopened cones. These flavor the bird's flesh with resin, making them unsuitable for human consumption. Though they are not likely to harm more established, mature pine forests, capercaillie can cause extensive damage to newly established plantations.

Other trees, including larch, spruce and occasionally some deciduous species, and herbaceous plants, make up the rest of the capercaillies' winter food. Because they feed in trees the birds are little affected by snowfall.

Displays at dawn

In April and May, usually at daybreak, cock capercaillie gather in a tree, on a rock or on the ground, to display. Like other members of the grouse family, they perform an elaborate ceremonial. Each male defends a small part of the display area, which is known as a lek. The neck is stretched up with the bill pointed to the sky so that the beard is well displayed, and the tail is fanned and held vertically like a peacock's tail. While in this position, the birds strut about rustling their wings and occasionally leaping as high as 3 feet (90 cm) into the air.

At the same time the capercaillie make various calls. One call has been described as sounding like cats fighting in the distance. Others are made up of noises which have been compared to

steel on a whetstone and corks being drawn from bottles. These sounds can still be called a song as they have the same function of establishing territory and attracting a mate as the songs of the familiar garden birds. During the breeding season, male capercaillie become highly aggressive and have been known to fly at humans who enter their home range, sometimes with sufficient force to knock the intruder to the ground.

The hens are attracted to the males' strutting performances and song displays. They mate with one of the cocks, then depart to raise families on their own. The nest, usually a hollow scraped in the ground, is made in the undergrowth in forests or in heather in open country. In the second half of April, 7 to 11, occasionally 5 to 16, eggs are laid. Until the clutch is completed, the eggs are covered with vegetation such as pine needles, after which incubation begins, continuing for 24–26 days.

Because incubation begins when the clutch is complete, the eggs develop together and hatch within 24 hours of each other. The chicks leave the nest when very young and are tended by the hen for some time. She protects them, luring away predators, such as foxes, by feigning injury. By limping away from the brood with one wing trailing as if broken, the hen attracts the attention of the predator, which chases what appears to be an easy prey. Once the brood is safe, the hen flies up, leaving the predator baffled. However, sometimes foxes use distraction displays themselves as a means of searching for chicks.

CAPUCHIN MONKEY

THERE ARE FOUR SPECIES of capuchin monkey, of which 33 subspecies have been recognized. The four species fall into two natural groupings, tufted and untufted. The first group has only one species, the tufted capuchin, *Cebus apella*. It has tufts of hair over the eyes or along the side of the head and a uniform coat of grayish brown hair. The other three species are the wedge-capped capuchin, *C. olivaceous*, the white-faced capuchin, *C. capucinus*, and the white-fronted capuchin, *C. albifrons*. The members of this group lack tufts on the head and have a patterned coat with patches of white on the face, throat and chest.

Capuchins are small monkeys, with the head and body measuring only 12–24 inches (30–60 cm); the tail is a similar length, reaching 2 feet (60 cm). The tail is prehensile and helps the monkeys maintain their balance in the trees. Capuchins are arboreal (tree-dwelling) creatures; they sleep on tree branches and descend to the ground only when they need to drink, or make short journeys on foot across open areas.

The name capuchin is derived from the resemblance of the hair on the head to the pointed cowl or capuche of Franciscan monks. The capuchins are also known as ringtail monkeys from their habit of carrying their tail with the tip coiled up. Capuchins live in the forests of South America. They range from Costa Rica south to Argentina and are also found on Trinidad. They have a life expectancy of 35 years, although those in captivity may live for longer.

Marching orders

Untufted capuchins live in groups known as troops, ranging from a small family party to a loose group of up to 50 monkeys led by one alpha (dominant) male. The groupings of tufted capuchins are generally smaller, numbering 5 to 15 individuals. Individuals transfer between troops after 2–5 years of age. Each troop keeps to regular tracks through the forest and has a small home range, which may overlap the ranges of other troops. Sometimes several troops may share a track, each using it at different times.

Young capuchins ride on their mothers' backs during early infancy. These are white-faced capuchins.

The capuchins (white-fronted capuchin, above) are small, agile monkeys that move swiftly through the forest canopy.

CAPUCHIN MONKEYS

CLASS	**Mammalia**
ORDER	**Primates**
FAMILY	**Cebidae**

GENUS AND SPECIES **Wedge-capped capuchin,** ***Cebus olivaceous*; white-faced capuchin,** ***C. capucinus*; white-fronted capuchin,** ***C. albifrons*; weeping (tufted) capuchin,** ***C. apella***

ALTERNATIVE NAMES
Sapajou; ring-tail monkey

LENGTH
Head and body: 12–24 in. (30–60 cm); tail: 12–24 in. (30–60 cm)

DISTINCTIVE FEATURES
Tip of long tail often coiled into ring. *C. apella*: tufts on head and above eyes; coat grayish brown. Other species: tufts absent; coat grayish brown, with white patches on face, throat or chest.

DIET
Mainly fruits, berries, nuts and invertebrates; some small vertebrates, such as young birds and lizards

BREEDING
Age at first breeding: 8 years (male), 4 years (female); breeding season: all year; number of young: 1; gestation period: about 180 days

LIFE SPAN
Usually up to 35 years

HABITAT
Tropical forests, from coastal mangrove swamps and mature rain forest to high-altitude dry forest

DISTRIBUTION
Central and South America; Trinidad

STATUS
Abundant; some subspecies vulnerable

Capuchin monkeys

The troop moves through the trees in single file and in a set order. First come the half-grown young of both sexes, then the adult females and the adult males; the females with young bring up the rear. Capuchins move by walking on all fours or by leaping. Adults are capable of making leaps of about 10 feet (3 m).

Tests have shown that capuchin monkeys can use insight to work out problems, rather than use laborious trial-and-error attempts at solution. Capuchins in captivity will use sticks to draw food toward the bars of their cages and are able to obtain fruit suspended out of reach by moving a box under it and climbing up.

Diet of fruit and insects

Capuchin monkeys are omnivorous, although fruit is their main food and they will raid orange, maize and cocoa plantations. Capuchins test fruit for ripeness before eating by squeezing, biting or smelling it. They eat no leaves or shoots. Insects, especially butterflies, are caught on the wing. Small birds and mammals are also eaten. Spiders and grubs are collected by prizing up the bark of trees. Hard fruit, nuts, beetles and birds' eggs are beaten against branches until soft or split open before being eaten. *C. capuchinus*, the only capuchin that is native to Costa Rica, will also eat nestling squirrels, young birds and small anolis lizards.

It is possible that capuchins aid seed dispersal for some tree species. In experiments scientists observed that two-thirds of the seeds which pass through a capuchin's gut germinate successfully. Moreover, these defecated seeds germinated 10 days earlier than uneaten seeds.

Fast developers

Tufted capuchin females will usually try to mate with the alpha (dominant) male of their troop, while untufted females will mate with any male in their group. Females have one baby at a time, born after about 180 days' gestation. At birth, the infant clings on to its mother's chest. After a short period of time, it switches to its mother's back or shoulders, hanging on to her fur with its hands and feet and wrapping its tail around her body. It leaves this position only at feeding times. After a while the baby starts to crawl around its mother's back and to examine objects. It then begins to move away from her for short distances, while she holds it by the tail in case it slips and falls. A few months later the baby becomes much more independent and its mother will push it away if it tries to climb on her. Capuchins develop very rapidly and are nearly full grown at 6 months. During this period, the father helps the mother to care for the baby; if an infant becomes separated from its parents and cries, other members of the troop will come to its assistance and bring it back to the safety of the troop. Young capuchins are usually fully weaned within 1 year.

Capuchin monkeys are preyed on by jaguars and large birds of prey, but they are rarely hunted by humans because their flesh is not as good to eat as that of the spider monkeys and woolly monkeys. The various species of capuchin are plentiful in the wild, although a few subspecies are vulnerable, due to the loss of their habitat and some localized hunting.

Young capuchins of a similar age make boisterous playmates. The weeping or tufted capuchin (below) is the only species with tufts on the head.

CAPYBARA

Despite their short legs and heavy bodies, capybaras can move quickly over land and are strong swimmers.

THE WORLD'S LARGEST RODENT IS the capybara. Also called water pig or water cavy, it resembles a huge guinea pig, or cavy, and rarely strays far from water.

The capybara (the name means "master of the grass") ranges over South America east of the Andes, from Panama to northeastern Argentina. Its body is massive, the head proportionately large and the legs fairly short. The overall length of the capybara is 40–50 inches (100–130 cm) and the height at the shoulder 20 inches (50 cm). When fully grown, an adult capybara may weigh up to 175 pounds (80 kg).

The capybara's coat of long coarse hair is sparse enough for the skin to be visible beneath it, which makes the skin very vulnerable to overheating and cracking in the sun. During the heat of the day, the capybara avoids sunburn by seeking shade or taking to the water. The giant rodent's coat is colored gray to reddish brown, yellowish brown on the underparts, with some areas of black on the face, the outer surfaces of the limbs and the rump.

The broad and deep head has small rounded ears, small eyes and nostrils, all set high up. Each of the front feet has four toes with hooflike claws, and there are three similar toes on each hind foot. All feet are slightly webbed. Capybaras have hard teeth; their incisors are orange-colored and can easily cut through even the toughest grasses and water reeds. The capybara often spends time gnawing on hard materials such as wood and nuts.

The male has a large whitish gland just in front of the eyes, known as the morrillo gland (morrillo is Spanish for hillock). During mating the male rubs the white oily liquid produced by the gland onto plants and then transfers the smeared liquid onto his body; the scent that the liquid gives off signifies the male's authority.

Strong swimmers

When alarmed, capybaras are capable of moving swiftly over land, galloping as horses do. They are also strong swimmers. Capybaras sometimes live in groups of up to 20, although the family

CAPYBARA

CLASS **Mammalia**

ORDER **Rodentia**

FAMILY **Hydrochaeridae**

GENUS AND SPECIES *Hydrochaeris hydrochaeris*

ALTERNATIVE NAMES
Water pig; water cavy

WEIGHT
60–175 lb. (27–80 kg)

LENGTH
Head and body: 40–50 in. (100–130 cm); shoulder height: 20 in. (50 cm); female larger than male

DISTINCTIVE FEATURES
Resembles large guinea pig. Stocky body; large, blunt muzzle; shaggy, reddish brown coat; webbed feet.

DIET
Aquatic plants and grasses near water bodies; occasionally crops such as grains, melons and squashes

BREEDING
Age at first breeding: 15–18 months; number of young: average 5; gestation period: approximately 150 days; breeding interval: usually 1 year

LIFE SPAN
Up to 8–10 years

HABITAT
Marshes, swamps, rivers and lakes with dense vegetation

DISTRIBUTION
Panama, Venezuela and Colombia south to Argentina

STATUS
Generally common but has declined in parts of range due to hunting pressure and habitat loss

Capybara

group is the basic unit, and favor dense vegetation in marshes and swamps, or wooded areas around rivers and lakes. One male is dominant in the group. Capybaras communicate through a series of barks, whistles and grunts, though they also have a well-developed sense of smell.

Normally capybaras swim with little more than ears, eyes and nostrils showing above the water, to enable them to breathe and to sense danger. They can swim considerable distances underwater if necessary, coming up among water plants and exposing only the nostrils. Rather than digging burrows like other rodents, capybaras scrape shallow beds in the earth.

Infant capybaras stay with their mother for up to 6 weeks, although other females will also nurse them.

Vegetarian feeders

Capybaras are generally crepuscular (active during the twilight hours) though they may become wholly nocturnal in places where they are regularly persecuted. Wholly vegetarian, capybaras feed on aquatic plants, often standing belly deep in water to do so. They also feed on grass and will sometimes graze with cattle. Occasionally they attack crops of cereals and fruits, such as melons and squashes, and as a result are culled as pests. Humans also hunt capybaras for their meat and fur and for their skins, which are used to make high-quality leather. Jaguars and pumas

Water hyacinths form a thick mat on the surface of pools, providing capybaras with both food and cover.

hunt capybaras on land and the rodents also fall prey to alligators and caimans; anacondas will sometimes pursue capybaras into the water.

Well-developed babies

Capybaras mate all year round, though mating is most common in April and May, just before the rainy season. Mating takes place in the water. After a gestation period of approximately 150 days the young are born on dry land as a single litter numbering two to eight, though the usual size is five. One litter is born per year.

Baby capybaras, born at an advanced stage of development, as in their relative the guinea pig, weigh about 3⅓ pounds (1.5 kg) at birth. They can see and walk, and are able to browse plants within a week. The young are very vocal, purring almost constantly. They remain with their mother until well toward the next breeding season, although other nursing females in the mother's pack also help to suckle them. The young are weaned in about 16 weeks and reach sexual maturity after 15 months, shortly after which they usually find a mate. The life span of capybaras is 8–10 years in the wild, though some have lived to 12 years in captivity.

Neutral buoyancy

The capybara looks at first sight like a typical land animal. However, it needs water not only to provide a refuge from potential danger, but also to keep cool and to preserve its health. If compelled to live long away from water, a capybara suffers from dry skin which soon becomes ulcerated, and the animal experiences difficulty in mating and defecating.

This large rodent is a good swimmer, yet it has few of the usual adaptations for spending time in the water. Its toes are only partly webbed and its body is not streamlined, yet like the hippopotamus, which has a similar bulky build, it is a graceful and accomplished swimmer. This is due to the fact that volume for volume the capybara is only slightly heavier than water. The animal's bulky body contains plenty of fatty tissue, the natural buoyancy of which counterbalances the weight of the bony skeleton. Thus, a small force exerted by the limbs is sufficient to propel the animal underwater. Capybaras can swim underwater for several minutes if necessary without coming up for air; they can even sleep underwater, as long as their nose is above the water's surface.

CARACAL

ALSO KNOWN AS desert lynx, the caracal is regarded by most zoologists as a close relative of the serval, one of the medium-sized cats. At one time its tufted ears led many to regard it as a true lynx, though this view has subsequently been discredited. In captivity the caracal has the catlike habit of rubbing itself against the bars of its cage and purring. It is distinguished by its reddish fawn, short-haired coat, long, slender legs and large black ears with very long ear tufts.

The caracal's name derives originally from the Turkish word *karakal*, meaning "black ears." In conjunction with various facial expressions, the caracal uses its ears to signal information such as its intentions or status to other caracals. The lengthy tufts at the tips of the caracal's ears emphasize the movements of the ears and head, though they may also serve to enhance the animal's hearing. In common with other cats, the caracal miaows, hisses and spits according to mood, but it also utters a unique coughing call during breeding.

The caracal's head and body are 2–3 feet (60–90 cm) long, with a 9–12-inch (23–30-cm) tail, and it stands 15–20 inches (38–50 cm) at the shoulder. Caracals weigh up to 42 pounds (19 kg). Their range includes scrub, steppe, savanna light woodland and semidesert through much of Africa, though they avoid deserts. They are also found in southwest Asia, including Arabia to Afghanistan, and eastward through Sind, Punjab and Kutch to Uttar Pradesh. Though rarely seen, the caracal is not thought to be endangered.

Jumping caracals

Caracals avoid forests and keep to sparse bush and grassland or to hilly country with boulders. As they are found in a wide variety of arid habitats, caracals are opportunistic hunters and prey on a range of wildlife. They can climb jackal-proof fences to kill poultry and are powerful jumpers. A caracal can strike birds that are already airborne by rearing up on its long, muscular hind legs or by jumping off all fours as high as 6 feet (1.8 m) into the air and swiping with its paws. It will also climb trees to pounce on birds nesting in lower branches.

Caracals feed on a variety of animals up to the size of goats, young kudu and the smaller antelopes; they will sometimes take sheep, and because of this farmers regard them as pests. After a kill, caracals will sometimes conceal the larger prey under vegetation rather than eating it on the spot, for later consumption. Larger kills may provide sufficient food to last a caracal several days. The caracal will also prey on foxes, rodents such as jerboas and hares, monkeys, guinea fowl and francolins, in addition to smaller

The caracal's distinctive ear tufts are believed to act as "flags" that signal information such as the animal's mood, intentions and status.

Caracals favor stony outcrops and rocky crags. This preference is reflected in their Arabic name, which translates as rock cat.

CLASS	**Mammalia**
ORDER	**Carnivora**
FAMILY	**Felidae**
GENUS AND SPECIES	***Felis caracal***

ALTERNATIVE NAMES
Desert lynx; rock cat

LENGTH
Head and body: 2–3 ft. (60–90 cm); shoulder height: 15–20 in. (38–50 cm); tail: 9–12 in. (23–30 cm)

DISTINCTIVE FEATURES
Long, slender legs; large, black ears with very long tufts; short, reddish brown fur

DIET
Birds, rodents and other mammals up to the size of small antelope

BREEDING
Age at first breeding: 1–2 years; breeding season: all year; number of young: usually 3; gestation period: 69–78 days

LIFE SPAN
Up to 17 years in captivity

HABITAT
Dry country, woodland, savanna and scrub

DISTRIBUTION
Most of Africa and Arabian Peninsula east to northwestern India

STATUS
Not known, but probably not threatened

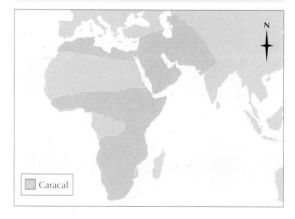

Caracal

birds. It is also reported to kill snakes and even to prey on the smaller African wildcat, where the ranges of the two species cross. Smaller prey are often taken up into the branches of trees, where the caracal is able to consume the kill in safety and at leisure.

Caracals are mainly nocturnal, so little is known of their habits. They will hunt by day in cool or cloudy weather, though when the sun is hot they prefer to rest in rocky crevices or in dense vegetation.

Underground nurseries

Caracals are solitary cats, though they will form pairs for mating and rearing young. They breed all year round, and after a gestation of 10–11 weeks, a litter of one to six, usually three, is born in an old aardvark burrow, foxhole or hollow tree. Young caracals are able to see after about 9 days, and at 1 month old they take meat for the first time.

Like any cat, the young are defended against intruders, the parent showing defiance by spitting like other members of the cat family. The kittens' first coat is bright reddish brown but after a few months silvery hairs grow, making the color more gray. Caracals develop quickly and are independent after 1 year.

With the exception of their distinctive tufted ears, caracals have the typical color of animals that dwell around the margins of deserts. Mammals living in arid or semidesert regions are generally colored pale buff to tawny or sandy red and freckled with gray. These colors harmo-nize with the color of the earth and surrounding landscape and so offer camouflage. Moreover, such coloring reflects heat and helps the animals to remain cool; no desert mammals are dark-colored. Some variation occurs in a caracal's coloring due to its age or region.

CARDINAL

A FAMILIAR GARDEN SONGBIRD of North America, the northern cardinal is also known as the red bird. Apart from a black "bib," the male plumage is a mixture of several shades of red, from deep scarlet to an intense reddish orange. The female's plumage is composed of a mixture of pinkish red and tinges of brown. Cardinals are finches, about 8⅔ inches (22 cm) long, with a stout bill which is nearly conical in shape, a conspicuous crest in both sexes and short, rounded wings.

The range of the northern cardinal extends from the temperate zone of the United States south to Mexico and Central America. It has also been introduced to southwestern California, Bermuda and Hawaii. In the latter, it breeds all year round and has become a pest because of the damage it does to fruit. Cardinals are also spreading unaided northward through the United States and are now a fairly common resident around New York and in parts of southern Canada. This spread may, at least in part, be due to the popularity of keeping bird feeders to attract birds into gardens and to provide them with a plentiful source of food in winter. Some cardinals migrate south in winter while others, especially the young birds, stay near the places where they were reared, in flocks of 6 to 24. It has been suggested that the practice of putting out food for birds may also be reducing migration.

The cardinals occur in a range of temperate and tropical habitats. On the arid Marias Islands off Mexico, the birds have to drink early morning dew before it evaporates, and cardinals have been seen drinking from pools formed at the bases of leaves. They prefer to live in open woodlands with clearings or in mixed growth at the edges of woods. Suburban gardens are very popular, providing both open ground and trees.

Accomplished songsters

Apart from their bright red plumage, cardinals are popular birds because of their song, which they perform nearly year-round. The male and female sing equally well so that it is difficult to distinguish the sexes by song alone.

In Tennessee the clear, whistling song of the male cardinal can be heard in January or February, when the ground is still snow-covered. The female starts singing during March; her song is a more muted version of the male's. Cardinal song can be heard throughout the summer and fall, occasionally as even late as November or December. The song is very varied. One male cardinal was recorded as producing 28 songs,

composed of different combinations of syllables. There is also a very quiet song, known as the subsong, which is mainly heard during the courtship season, in February–April.

The song plays an important part in courtship and nesting. As is usual when both sexes sing, both male and female defend the territory. The female drives out intruding females but ignores strange males, which are repelled by her mate. The males sing in order to attract females. Cardinals have a wide range of songs that vary from place to place and which they learn to sing in a particular way by imitating the songs of other cardinals around them. Females do not appear to favor males with one particular song, though it may be that the intensity and duration of the song is important in female choice. Strong singers must be strong birds because in terms of energy singing is very costly for the male cardinal; the equivalent of a human singing at the top of his or her lungs for a few hours. Cardinal males sometimes have two female mates. The nestlings of one female will be provisioned by the male, but the other female's brood will tend to be neglected.

Nest-building starts in March or April. Sometimes the male helps; at other times he accompanies the female while she flies from place to place collecting weeds, leaves, grasses

Female and young male cardinals have a predominantly brown plumage, enlivened by a pinkish red suffusion to the breast.

The male cardinal's gaudy plumage ensures its popularity with humans. By putting out food to attract the species, garden owners have helped it expand northward through the United States.

CARDINAL

CLASS	**Aves**
ORDER	**Passeriformes**
FAMILY	**Emberizidae**
GENUS AND SPECIES	*Cardinalis cardinalis*

ALTERNATIVE NAMES
Northern cardinal; red bird

LENGTH
Head to tail: 8⅔ in. (22 cm); wingspan: 9–12 in. (23–30 cm)

DISTINCTIVE FEATURES
Strong, conical bill; large crest. Male: bright red plumage with black bib. Female and young: brownish plumage, pinkish on breast.

DIET
Seeds, fruits and invertebrates

BREEDING
Breeding season: March–August; number of eggs: 3 or 4; incubation period: 12–13 days; fledging period: 10–11 days; breeding interval: 2 to 4 broods per year

LIFE SPAN
Not known

HABITAT
Woodland edge, thickets, scrub, gardens and swamps

DISTRIBUTION
Southeastern Canada and eastern U.S. south to Mexico, Guatemala and Belize

STATUS
Common or abundant

Cardinal

and rootlets, which are then woven into a bowl. The nest itself may be compact and well-lined or a very loose, flimsy structure. It is usually situated in a shady location, such as in a hedge or among the branches of a young evergreen. However, the site of the nest can vary considerably; some are built on the ground, while others are placed high in a tree.

The first clutch is started shortly after the nest is completed, and consists of three or four eggs, though five have been recorded. Clutches laid later in the season usually have only two eggs. Incubation takes 12–13 days, starting when the last egg is laid. The eggs are pale blue in color with brown and pale purple speckles. The chicks leave the nest when 10–11 days old. During this time both parents feed them. As is often the case with fruit-eating birds, at the outset the young are fed on insects, which provide a high-protein diet necessary for rapid growth.

Multiple broods

The cardinal raises two to four broods during the year. In Tennessee, for example, breeding continues from April to August. Further south, breeding may take place year-round. The pair stays together throughout the breeding season and may keep company during the winter. While the female builds a new nest and incubates the next clutch, her mate continues to feed the previous brood, which are finally chased out of the territory when the next brood is hatched.

Young cardinals begin to sing when they are 3–6 weeks old. At first the song is warbling, not at all like the adult song, which is a much-repeated whistle of varying pitch. However, adult phrases are added by the age of 2 months and the full song develops by the next spring.

CARDINAL FISH

CARDINAL FISH IS A NAME given to nearly 200 species of fish in about 23 genera. The fish are usually red or have red in the pattern of the body, and live mainly on coral reefs in tropical seas or lagoons. A few live in freshwater streams in the tropical areas of the Pacific, Atlantic and Indian oceans. Still others live in temperate seas.

Cardinal fish have a distinctive body shape, with a large mouth, short and spiny dorsal fins which are fully separated and a long caudal peduncle. Apart from a few deep-sea forms they are shallow-water fish, usually not more than 4 inches (10 cm) long. The largest are up to 8 inches (20 cm) long and live in the brackish water of mangrove swamps. Cardinal fish are often present in great numbers, suggesting that they are the mainstay of many predatory fish.

Shell shelters
The best-known cardinal fish are two species of 2-inch (5 cm) long conchfish found in Florida, Bermuda and the Caribbean. These shelter inside the large mollusks of the genus *Strombus*, known as conch shells, resting in the mantle cavity and coming out at night to feed on small crustaceans.

Other cardinal fish shelter inside sponges and empty bivalve shells, or in any other convenient hollow object or cavity.

Spiny home
One cardinal fish, *Siphamia versicolor*, of the Nicobar Islands in the Indian Ocean, lives in association with a dark red sea urchin. When undisturbed, the urchin parts its long spines so that they form pyramidlike clusters and the fish moves between these, cleaning the sea urchin's skin. At the slightest alarm, even a shadow falling on the urchin, the spines are spread defensively and the cardinal fish shelters among them, usually with its head downward. At night the fish comes out to feed, but if it is driven from its original sea urchin host, it will swim to another. If this new urchin is of a different color, the fish will change its own coloration to match that of its new host.

Mouth nurseries
As the female cardinal fish lays her eggs, the male takes them into his mouth and there they remain until they hatch. This is known as mouth brooding. Sometimes the males alone

Cardinal fish favor warm and tropical waters. These orange-lined cardinal fish, Archamia fucata, live in the Red Sea.

389

CARDINAL FISH

CLASS	**Osteichthyes**
ORDER	**Perciformes**
FAMILY	**Apogonidae**
GENUS	**About 23, including *Apogon*, *Archamia*, *Siphamia*, *Sphaeramia***
SPECIES	**Almost 200, including *Apogon imberbis* (discussed below)**

LENGTH
4–8 in. (10–20 cm)

DISTINCTIVE FEATURES
Small, compressed body; large eyes; reddish overall, darker above; base of caudal fin has 2 or 3 dark spots

DIET
Small invertebrates and fish

BREEDING
Breeding season: June–September; eggs mouth brooded by male; in other species eggs mouth brooded by female

LIFE SPAN
Not known

HABITAT
Seas with rocky beds at up to almost 650 ft. (200 m) deep; in summer in shallows

DISTRIBUTION
Mediterranean and eastern Atlantic from Portugal to Morocco and Azores. Other species range southward to southern Africa, Gulf of Guinea, Indian Ocean and Pacific.

STATUS
Common

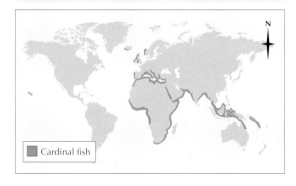

Cardinal fish

The coral cardinal fish, Sphaeramia nematoptera, *has the separated dorsal fin, large eyes and bright colors characteristic of its family.*

hold the eggs, sometimes the females. There are a few species in which the eggs are picked up by the male only when danger threatens.

The female of one species of cardinal fish found in Australian waters lays eggs ¼ inch (0.5 cm) in diameter and the male can hold 150 in his mouth. In one Mediterranean species the eggs are only 0.5 millimeters in diameter and the male can hold 22,000 at one time. The eggs are fertilized inside the female's body and there is a curious inversion in the copulation procedure. The genital papilla of the female is long and she inserts it into the male's body to receive the sperm for fertilizing her eggs.

Little is known about the enemies of cardinal fish. Consequently, it is possible only to presume that they must be eaten by many small and medium-sized predatory fish. This is supported by the behavior of the brownspot cardinal fish. The 3-inch (7.5-cm) brownspot, so named because of a dark spot above each of its breast fins, flops on one side and gives the appearance of being dead when an attempt is made to catch it.

Light organs

Most cardinal fish are nocturnal and hover in caves or crevices during the day. Some species have light-emitting organs. One species of cardinal fish, *Apogon ellioti*, from Southeast Asian seas, contains luminous bacteria in its throat. The gland in which these bacteria are located has a reflector and the muscles below the gland are translucent and act as a lens. In other cardinal fish these light-giving bacteria are found in the gut. Some cardinal fish, such as those in the genus *Rhabdamia*, have transparent bodies to facilitate the spreading of light from their internal organs.

CARIBOU

CARIBOU AND REINDEER are now usually considered to be the same species. Animals found in northern Europe and Russia are generally called reindeer, while those that occur in North America and Siberia are called caribou. There are seven subspecies of caribou: barren-ground (*Rangifer tarandus granti*), Svalbard (*R. t. platyrhynchus*), European (*R. t. tarandus*), Finnish forest (*R. t. fennicus*), Greenland (*R. t. groenlandicus*), woodland (*R. t. caribou*) and Peary (*R. t. pearyi*). The species is plentiful in the wild but many domesticated herds also exist.

Caribou have longer legs than reindeer, standing 3½–4 feet (107–127 cm) at the shoulder and weighing up to 600 pounds (270 kg). The coat varies from almost black to nearly white, but most caribou are brownish or grayish with a light rump and underparts. In winter they become lighter, appearing more heavily built as fat is laid down, and the hair, consisting of woolly underfur and stiff guard hairs, lengthens. The ears and tail are short and the muzzle is unique, being furry. These adaptations reduce heat loss in cold weather.

Reindeer and caribou are the only members of the deer family in which both sexes bear antlers, although those of the females do not reach the large size attained by the males, which also have a ruff of long hair on the throat. Large bulls (males) shed the velvet (fur) covering their antlers in late August and early September as a precursor to the rutting (mating) season.

Mass migrations

Caribou live in small bands of 5 to 100 animals or in herds of up to 3,000. There is little organization and no established leader; the groups merely move together and close on each other when alarmed.

In April and May the herds migrate north to the open tundra, where they live until June or July, when there is a movement back to the wooded parts at the southern end of their range. The migration follows regular paths and the trails become cleared of vegetation and beaten flat over years of use. In September there is a second northward movement back to the tundra, though not as far as the summer feeding grounds, for the rut. After the rut the herds return to the woods for the winter but a few small bands remain in the tundra all winter.

The caribou's large feet help in its long migration over snow, slippery ice and sinking bogs. The two halves of the cloven hoof are very broad and splay out, reducing the pressure on the ground, thereby acting like snowshoes. The pressure exerted on the ground is 2 pounds per square inch (0.15 kg/cm^2), a very low figure. By way of contrast, the moose exerts a pressure of 8 pounds per square inch (0.55 kg/cm^2).

Unlike in other members of the deer family, both sexes of caribou grow antlers.

CARIBOU

CLASS	**Mammalia**
ORDER	**Artiodactyla**
FAMILY	**Cervidae**
GENUS AND SPECIES	**_Rangifer tarandus_**

ALTERNATIVE NAME
Reindeer

WEIGHT
200–600 lb. (90–270 kg)

LENGTH
**Head and body: 4¼–6¾ ft. (1.3–2.1 m);
shoulder height: 3½–4 ft (107–127 cm);
male larger than female**

DISTINCTIVE FEATURES
**Thick muzzle with furred nose; maned neck;
broad hooves; large antlers (both sexes)**

DIET
**Summer: leaves, shoots, buds, grasses,
sedges and mushrooms; fall and winter:
lichens, dried grasses and twigs**

BREEDING
**Breeding season: September–mid-October;
number of young: 1 or 2; gestation period:
225–240 days; breeding interval: 1 year**

LIFE SPAN
Up to 10 years

HABITAT
**Tundra and mountain slopes; also open
woodland in winter**

DISTRIBUTION
**Alaska, northern Canada and parts of
Greenland; Arctic Europe, Siberia and
northern Asia**

STATUS
**Population: about 5 million worldwide.
Caribou: approximately 1,000,000 in
North America. European reindeer:
25,000 in Scandinavia and 20,000 in
northern Russia.**

Caribou

*Caribou are strong
swimmers and rivers
are no obstacle to
their migrations.*

The concave shape of caribou hooves and the
patches of hair on their underside give the
animal a good grip on slippery surfaces. During
migration the herds move forward at the rate of
about 20 miles (30 km) a day, but when hard
pressed, caribou can gallop short distances at
speeds of over 40 miles per hour (65 km/h).

Lichen feeders

In winter, caribou eat lichen—the so-called
reindeer moss—and dried grasses, which they
reach by scraping away the snow with their
hooves. They also browse on willow and aspen
twigs. In summer they have a broader diet that
includes a number of trees and plants, such as
birch, willow, horsetails, grasses and sedges.
Every year the antlers are shed and these cast

antlers are chewed, helping to rebuild the body's supply of calcium salts in preparation for the growth of new antlers.

Mating season

The annual caribou rut takes place in September–October. Unusually for members of the deer family, the bulls serve the cows (adult females) indiscriminately, without forming harems. Aggression occurs only when the bulls come near one another, and any sparring is short-lived. During the rut, the bulls thrash the undergrowth with a side-to-side movement of their antlers. This is part of the courtship display and it has been suggested that the enlarged brow-tines (projected parts) on the antlers of a bull caribou are an adaptation to protect the eyes while thrashing the stiff, woody stems of bushes.

Well-timed births

The young caribou are born in early June, while the herds are on their spring migration. Ninety percent of all births occur within a 2-week period. If the young are born too early, they will succumb to bad weather; those born too late will not have time to develop sufficiently to withstand the coming winter. As the herds are migrating, it is an advantage for the cows to drop their young at the same time so that the animals all move at the same speed and there is less danger of stragglers being left behind, where gray wolves are waiting to make an easy kill.

The calves weigh 9 pounds (4 kg) when born and can run when half an hour old; in 4 hours they can outrun a human. Such fast development is vital, as the mother and calf must keep up with the rest of the herd. To encourage the calf to follow her, the cow faces the calf and bobs her head, grunting at the same time. This stimulates the calf to follow.

Generally the cows sort themselves into small bands, each made up of cows with calves of roughly the same age, so that they can run together at the same speed. When very young, the calves will follow any moving object, so when the band is disturbed, the calves follow blindly. When the danger is past, their mothers sort them out, identifying their own calves by scent. By the end of their first month the calves begin to graze, but they will continue to suckle throughout the winter.

Brown bears sometimes prey on young caribou; otherwise gray wolves are the main predator. Groups of wolves associate with the caribou herds for most of the year, preying on sick, old or very young animals, or any that become separated from the herd. It has been calculated that not more than 5 percent of a given caribou population succumbs to wolves.

Wolves have lived around caribou herds for tens of thousands of years. By killing and eating the weaker caribou, wolves actually strengthen the herd as a whole.

Provider of all

The caribou were at one time the northern equivalent of the bison of the North American plains, providing humans with nearly all their basic needs. Both species once existed in vast herds that represented an apparently endless source of food, clothing, housing material and raw material for many implements.

Caribou bulls shed the velvet on their antlers in late August and early September in preparation for the rutting (mating) season.

Caribou are among the most migratory of all mammals. Herds still follow traditional routes that have been used for decades.

Northern Native Americans and Inuit people based their culture around the caribou, as did the inhabitants of northern Scandinavia and northern Russia. Native Americans knew the migration trails of the caribou and would wait for the herds to pass, drive them into corrals and slaughter them. The hunters would also attack the herds as they crossed rivers, intercepting them with canoes and spearing them. The nomadic lifestyle of some tribes was based on the migrations of the caribou herds.

Reindeer and caribou were used to supply milk, cheese and meat. Their sinews were used for sewing clothing or for covering canoes while their bones and antlers were used to make needles and other tools. The fat from caribou and reindeer provided fuel and light, and their skins were used to make lightweight, waterproof clothing. The animals' ability to resist the cold and to forage for themselves meant that they did not require stabling and made them ideal to draw sleighs or to act as pack animals. However, the increased use of manufactured goods and processed foods has meant that reindeer are now used less frequently as sources of food and as beasts of burden.

Hunters using traditional Native-American weapons killed only limited numbers of caribou and did not threaten the species' survival. However, the introduction of new technology such as rifles by European settlers dramatically reduced the populations of caribou in North America. The decrease in caribou numbers indirectly devastated the lifestyle of the Native-American tribes that relied on caribou herds.

Survival of the species

In the early 20th century an estimated 1.75 million caribou roamed the barren grounds of northern Canada, between the Mackenzie River and Hudson Bay, from Victoria Island in the north almost to Lake Winnipeg in the south. This is a region of generally flat country with many lakes and bogs. The caribou population is now much reduced due to a combination of overhunting, destruction of food by forest fires and high mortality in severe weather. Having dropped to about 279,000 in 1955, the population has since risen to around 1 million through an intensive program of conservation, including the introduction of enforced gaming regulations and supervised hunts.

CARP

The carp has an acute sense of hearing. Its large swim bladder is divided into two or three segments and is connected via modified vertebrae to liquid-filled chambers in the inner ears of the fish. The first four vertebrae behind the skull act as a series of levers, transmitting sound waves picked up by the swim bladder to the middle ear. This connecting system is called the Webnerian mechanism and is also present in catfish and characin.

Wild carp at home

Carp prefer shallow, sunny waters with a muddy bottom and abundant aquatic plants. They avoid clear, swift-flowing, shaded or cold waters. Today wild carp are mainly found in large rivers. They feed on insect larvae, freshwater shrimps and other crustaceans, worms and snails, as well as some plant matter. The barbels (organs of touch) and the protrusible mouth are used for grubbing in the mud, much of which is swallowed and later ejected when the edible parts have been digested. In winter feeding ceases and the fish enter a resting period, a form of hibernation.

In May to July carp move into shallow water to spawn, laying their eggs on the leaves of water plants. Each female lays over 60,000 eggs per pound (about 120,000 eggs per kg) of her body weight. After the larvae hatch out, in 5–6 days, the adults returning to deeper water, while the young fish remain in shallow water, near the bank. The males become sexually mature in 3–4 years; the females mature in 4–5 years. Small carp will be eaten by almost any fish significantly larger than themselves, including larger carp.

As with many other domesticated species, carp are found in a number of varieties, of two main types: leather carp and mirror carp. The first is scaleless; the second has large scales in two rows on each side of the body. The shape of the body varies from relatively slender to deep-bodied with a humpback. Carp have probably been domesticated for many centuries, and have been transported all over the world for ornamental ponds or for food. Their large size, quick growth and ability to withstand high temperatures has made them a popular fish for farming. The commercial breeding of carp is a

The carp uses its protrusible mouth and barbels to stir up mud in search of food.

O F THE EXTENSIVE CYPRINIDAE family, the carp, *Cyprinus carpio*, is the most widely distributed. Native to Japan, Central Asia from Turkestan to the Black Sea, and China (where many exotic varieties were bred), it has also been introduced into Africa, Australia and many European countries. It is believed that the carp were first introduced to the British Isles in the 1300s and to the United States in the 1870s.

The carp differs from other members of the Cyprinidae family in its long dorsal fin, with 17 to 22 branched rays, the strongly serrated third spine of the dorsal and anal fins, and in its four barbels, two at each corner of the slightly protrusible (extendable) mouth. There are no teeth in the mouth, but there are throat-teeth. Wild carp are olive to yellow green on their backs, greenish yellow to bronze yellow on their flanks, and have yellowish underparts. Their fins are gray green to brown, sometimes reddish. Ornamental varieties of carp are known as koi.

Chemical communication

Specialized cells in the skin of carp contain chemicals known as pheromones, which act as a defense mechanism. If the fish is injured, the pheromones are released into the surrounding water. Other fish that detect the chemicals rapidly become agitated. They scatter and hide rather than attack the wounded fish.

There are more than 2,000 species of carp worldwide, including the Crucian carp, Carassius carassius.

COMMON CARP

CLASS	**Osteichthyes**
ORDER	**Cypriniformes**
FAMILY	**Cyprinidae**
GENUS AND SPECIES	***Cyprinus carpio***

ALTERNATIVE NAME
Minnow

WEIGHT
Up to 60 lb. (27 kg), usually much less

LENGTH
Up to 40 in. (102 cm)

DISTINCTIVE FEATURES
Four barbels (2 long and 2 short) on upper jaw; protrusible (extendable) mouth that lacks teeth; 1 to 3 rows of pharyngeal (throat) teeth; greenish brown to brown body, yellowish on lower parts

DIET
Insect larvae, small snails, shrimps and other crustaceans, worms and some vegetable matter

BREEDING
Age at first breeding: 3–4 years (male), 4–5 years (female); breeding season: May–July; number of eggs: about 2 million; hatching period: 5–6 days at 73–75° F (23–24° C)

LIFE SPAN
Up to 50 years, usually less than 15 years

HABITAT
Slow or stagnant, well-vegetated reaches of lakes and rivers

DISTRIBUTION
Native to Japan, China and Central Asia; widely introduced in North America, southern Africa, New Zealand, Australia and Eurasia

STATUS
Common, but declining in many areas

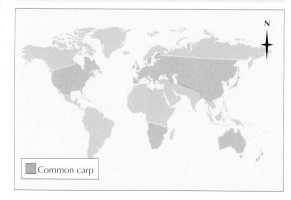

Common carp

major industry for many countries, especially those in eastern Europe and Israel. The carp is a resilient fish. It can survive for lengthy periods and, when removed from water and wrapped in damp moss or water plants, is able to survive long-distance travel. The carp is also resistant to low oxygen levels and high temperatures and can survive in contaminated water that would prove fatal to other fish. A carp can survive for weeks during dry periods by burrowing into mud.

The curious carp

There is a surprising conflict of opinion among scientists on important points such as the carp's longevity and maximum weight. Tate Regan, an authority on fish in Britain during the first half of the 20th century, was of the opinion that under artificial conditions a carp might attain 50 years of age but that 15 years would probably be the maximum age in the wild.

Carp usually grow to about 25 pounds (11 kg) in Britain but on the continent a fish of over 60 pounds (27 kg) and a length of 40 inches (102 cm) has been recorded. Claims have been made for 400-pound (181-kg) carp. Frederick II of Prussia is said to have caught one of 76 pounds (34 kg) and a 140-pound (64-kg) carp is said to have been caught at Frankfurt on Oder, Germany.

Although it is difficult to assess the population size of many fish, Cyprinidae is believed to be the most endangered fish family, with over 150 species listed as threatened. It has been estimated that the world population of wild common carp is likely to decline by at least 80 percent over the next 10 years, due to shrinking habitats, pollution, parasites and other threats.

CASSOWARY

THESE LARGE FLIGHTLESS birds live in dense rain forests on New Guinea, its adjacent islands and the northeastern seaboard of Queensland, Australia. Their drooping black plumage is made of coarse, bristlelike feathers. The skin of the neck and head is naked and brightly colored. It is also ornamented with colored wattles except in the dwarf cassowary, the colors being red, blue, purple, green and yellow. The head is crowned by a helmet or casque. This was once thought to be a bony extension of the skull but is now known to be composed of a tough, foamlike substance. The long, strong legs bear three toes, the innermost of which has a particularly long stout claw. The word cassowary derives from *kesuari*, a Malay name for the birds.

The wings are very small and hidden. Owing to the length of the aftershaft, the small tuft at the base of the vane, the feathers appear to be double. The quills of the wing feathers are without vanes. But the shafts of the wing feathers remain as horny spines up to 15 inches (38 cm) long. There are three species of cassowaries: the southern (*Casuarius casuarius*), northern (*C. unappendiculatus*) and dwarf (*C. bennetti*) cassowaries. The largest species, the southern cassowary, grows up to 5½ feet (1.7 m) in height and weighs up to 128 pounds (58 kg).

Fast forest runners

Cassowaries are secretive birds, keeping to dense jungle and rapidly making for cover if surprised in the open. They are more often heard than seen, their call being a deep booming or croaking. They can run at 30 miles per hour (48 km/h) even through dense thorny undergrowth, with the head down and forward, protected by the tough casque. The stout wing quills, held out and curving to the line of the body, ward off thorns and entangling vines. However, they tend to follow regular tunnel-like runs through the jungle vegetation. Cassowaries can leap obstacles, plunge into water and swim rivers, and in defense can leap high to make raking blows with their long daggerlike claws.

The most studied of the three species, the southern or Australian cassowary, goes about singly or in pairs, and occasionally up to six may be seen together. The birds rest during the hottest part of the day in sunny places, which they visit regularly. The food of all three species is mainly fallen fruit, especially palm seeds, plums and figs, but also includes some fungi, invertebrates and carrion. Fruit is sometimes taken directly from low bushes.

Caring father

The male is smaller than the female, but he is responsible for the care of the offspring. The female lays three to six greenish eggs, each about 5 inches (13 cm) long, on a nest of leaves at the base of a tree, during June–August.

The cassowary's casque is made of a foamlike substance. When the cassowary is running through thick undergrowth in the rain forest, the casque protects the bird's head.

A male southern cassowary guarding his clutch of eggs. Male cassowaries take on all parental responsibilities.

The male incubates them and guards the chicks for a further 7 weeks. During this time, if disturbed, he races away in what is thought to be an effort to divert the intruder from the nest. The adult male is conspicuous; the eggs are greenish and blend in with leaves and other vegetation. The chicks are buff with brown stripes.

Hunted by humans

Humans have hunted cassowaries for their flesh and feathers for centuries. New Guinea aborigines also keep the chicks as pets. As the chicks grow older they are placed in crude, cramped cages. Some are kept for their plumes, which are regularly plucked; others are eaten when they are full-grown. Land clearance for tourism and residential development has had a far more damaging impact on cassowaries, however. It is estimated that there are no more than 1,500 to 3,000 southern cassowaries in existence today. About 40 individuals are currently living in captivity in Australia but the future of this species is by no means assured.

SOUTHERN CASSOWARY

CLASS	**Aves**
ORDER	**Casuariiformes**
FAMILY	**Casuariidae**
GENUS AND SPECIES	***Casuarius casuarius***

ALTERNATIVE NAMES
Australian cassowary; double-wattled cassowary; two-wattled cassowary

WEIGHT
64–128 lb. (29–58 kg)

LENGTH
Head and body: 4–5½ ft. (1.3–1.7 m)

DISTINCTIVE FEATURES
Very large body; long, powerful legs; small wings are mainly hidden; blackish body plumage; naked blue neck with two red wattles; large casque on head

DIET
Mainly fallen fruits; also invertebrates, fungi and carrion

BREEDING
Age at first breeding: 30–42 months; breeding season: eggs laid June–October; number of eggs: 3 to 5; incubation period: 49–56 days; breeding interval: 2 or 3 broods per year

LIFE SPAN
Usually 12–19 years

HABITAT
Tropical rain forest and swamp forest, especially near forest edge, clearings and streams

DISTRIBUTION
New Guinea and adjacent islands; northeastern Queensland, Australia

STATUS
Vulnerable; estimated population: 1,500 to 3,000

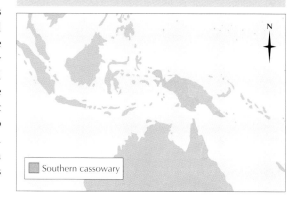

Southern cassowary

CATBIRD

THREE WIDELY DIFFERING groups of birds have been given the name catbird. The Australian catbirds are closely related to the bowerbirds. One of these, the tooth-billed catbird or stagemaker, *Ailuroedus dentirostris*, clears an arena on the ground and covers it with fresh leaves, cut with the serrated edge of its bill. The Abyssinian catbird, *Parophasma galinieri*, is regarded by some authorities as a species of flycatcher and by others as a babbler.

The North American gray catbird, *Dumetella carolinensis*, considered here, is one of the mockingbird family, and is a familiar garden bird. Its name is derived from the catlike mewing notes of its song. The gray catbird is up to 7 inches (18 cm) long, and has a wingspan of 9½–10 inches (24–26 cm). It is distinguished by its black cap and by the russet patch under the base of its tail. In the same family there is the black catbird, *Melanoptila glabrirostris*, of Yucatan in Mexico, which has brilliant iridescent black plumage.

The gray catbird breeds from southern Canada through the Midwest and eastern states to the Gulf of Mexico. Part of the population is resident year-round, but most catbirds migrate southward in autumn to Central America and the Caribbean. They do not cross the Gulf of Mexico in their migration because they do not have the fat reserves to do so, and instead follow the coast. Gray catbirds also breed in Bermuda.

Life on the edge

In countryside untouched by humans, gray catbirds nest in undergrowth at the edges of forests, along streams and marshes or in clumps of shrubs in open country. This preference for marginal habitats, where dense undergrowth borders open space, must have limited the gray catbirds' range and numbers in the days before Europeans invaded the American countryside. Now that vast tracts have been opened up, leaving islands of woodland, the catbirds have been able to spread.

As the European settlers pushed west, logging and road-building opened up the forests. The dramatic increase in human population that followed brought first the hedgerows and copses associated with farmland, and then the parks and gardens of suburban towns where the catbirds could nest in shrubberies. Gray catbirds are now firmly established as garden birds and their

song and mimicking quickly endear them to householders. Although gray catbirds nest, and usually sing, from the cover of shrubberies, they easily become tame and are content to feed around houses.

A diet of insects and fruit

That catbirds are primarily insect-eaters is indicated by the thin, pointed bill, which is typical of insect-eating birds. Many different invertebrates are taken, including weevils, caterpillars, grasshoppers, beetles, bees and wasps. Catbirds, however, are also fruit-eaters and have sometimes been condemned as pests of grape and blackberry crops, although this is probably seldom a serious problem for farmers. Gray catbirds also feed on the fruits and berries of a variety of wild and cultivated shrubs, such as privet, alder and hawthorn. The introduction of cultivated shrubs such as Tartarian honeysuckle, lilac and forsythia has, along with the general cultivation and opening up of the land, helped the catbirds to increase in numbers, as these shrubs provide both food and nesting sites. The catbirds, in turn, have helped to spread these introduced species by scattering the seeds in their droppings.

Nesting in the undergrowth

Male gray catbirds start singing shortly after their return from their winter quarters and courtship and nesting start within a week of the songs. When the pair has formed, a nest site is chosen, usually 2–3 feet (60–90 cm) into the

The gray catbird is so named because of its distinctive call, which resembles a cat's meow.

Gray catbirds easily become tame and are a familiar sight in many gardens across North America.

GRAY CATBIRD

CLASS	**Aves**
ORDER	**Passeriformes**
FAMILY	**Mimidae**
GENUS AND SPECIES	***Dumetella carolinensis***

ALTERNATIVE NAMES
Common catbird; northern catbird

LENGTH
Head to tail: 7⅓ in. (18.5 cm); wingspan: 9½–10 in. (24–26 cm)

DISTINCTIVE FEATURES
Black crown; upperparts brownish gray, underparts slate gray; undertail area russet

DIET
Insects and berries

BREEDING
Age at first breeding: 1 year; number of eggs: usually 3 or 4; incubation period: 13 days; fledging period: 11 days; breeding interval: up to 3 broods per year

LIFE SPAN
Not known

HABITAT
Woodland edge, suburban gardens, parks and orchards

DISTRIBUTION
Breeding: southern Canada south to Gulf of Mexico; winter: southern states of U.S., Mexico, Central America and Caribbean

STATUS
Common

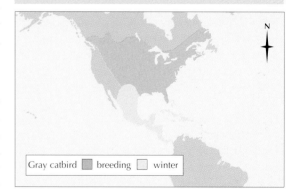

foliage of a shrub, but sometimes fairly exposed. The male leads the female in the search for a suitable nest site. Flying in front of her, with a piece of nest material in his bill, he flies to a fork in a shrub, singing all the time.

The female watches as he proceeds to play about with the scrap of material as if molding it into a nest, bowing and fluttering his wings at the same time. At this stage the foundations of a nest may be built, then abandoned, while the proper nest is built a short distance away and about 5–6 feet (1.5–1.8 m) from the ground. The nest is made of twigs, grasses, leaves and roots. The female does most of the work, the male merely bringing her some of the material.

The clutch consists of one to six eggs, usually three or four, and is incubated for about 13 days. The chicks spend about 11 days on the nest and are then taken some distance away by the parents. They are fed for a further 2 weeks. Depending on the weather, two or three broods may be raised in a season. The chicks are fed on insects or fruit. In the northern part of the range the first brood is fed only on insects alone; the second has the additional benefit of fruit, which has by then become available.

Nest raiders

The nests of gray catbirds are raided by many animals. Mammals including chickarees, chipmunks, mink and raccoons, and birds such as crows are the main predators in the wild. In suburban areas the nocturnal forays of the domestic cat probably do the most harm. Snakes also destroy gray catbird nests. Sometimes the parent catbirds will try to drive off the marauder by pecking and beating it with their wings.

Brown-headed cowbirds, *Molothrus ater*, which have the same parasitic habits as cuckoos, will attempt to lay their eggs in gray catbird nests. However, even if a cowbird manages to evade the parent catbirds' attacks and lay its eggs, these will later be thrown out. Moreover, gray catbirds will actually destroy the eggs of other songbirds. Experiments suggest that they do this not only to consume the eggs but also to reduce the risk of parasitism by cowbirds.

CATFISH

THE EUROPEAN CATFISH grows to 16½ feet (5 m) or more in the stagnant waters of central and eastern Europe and western Asia. Also known as the wels, it is the largest and most famous of a large group of species known as the naked catfish, which lack scales.

The head of the European catfish is large and broad, the mouth has a wide gape and around it are three pairs of barbels or "whiskers," the feature all catfish have in common. In the European catfish one of the three pairs of barbels is very long and highly mobile. The eyes are small and the body is stout, almost cylindrical in front, and flattened from side to side in the rear portion. The skin is slimy and has no scales. The fins are small, except for the long anal fin. Catfish are olive green in color, becoming darker on the back, the sides are mottled with whitish blotches, and the belly is whitish. The European catfish is also known as the sheatfish and as *silurus*, the name given it by the Romans.

Night hunter

The European catfish lives in sluggish rivers or stagnant water bodies with plenty of water plants. It spends the day under overhanging banks or on the mud in deep water, by night foraging in the silt with its barbels in search of small invertebrates. The catfish feeds voraciously on fish, crustaceans and frogs. The larger ones take small water birds and mammals. In May to July, the breeding season, the catfish moves into shallow water, where the female lays her eggs in a depression in the mud formed by lashing movements of her tail. A large female may lay 100,000 eggs, which are guarded by both parents. The fry are black and tadpole-shaped.

Variety of species

The naked catfish show great diversity in both form and habits. The banjo catfish of the family Aspredinidae live in rivers and brackish estuaries in tropical South America. They are named for their flattened bodies coupled with an unusually long tail. In one species of banjo catfish, *Aspredinichthys tibicen*, the tail is three times the length of the body. In the breeding season, the females of this species grow a patch of spongy tentacles on the abdomen and carry their eggs anchored to these. Another species of banjo catfish has evolved a form of jet propulsion, in which water is rapidly expelled from the opercular cavity (the cavity behind the gill coverings) to power the fish forward.

North America is home to a large family of catfish, the Ictaluridae, which has representatives from Canada south to Guatemala. The flathead catfish, *Pylodictis olivaris*, grows up to 5½ feet (1.7 m) long and may weigh as much as 100 pounds (45 kg). The channel catfish, *Ictalurus*

The "whiskers" of catfish are in fact tactile barbels used for locating food hidden in the mud. This is the sharptooth catfish, Clarias gariepinus.

Catfish range in length from the European catfish, which can exceed 16 feet (5 m), to the candirú, which measures only about 1 inch (2.5 cm).

punctatus, is a valuable food fish and is found from Mexico and Florida north to southern Canada. Another important food fish is the blue catfish, *I. furcatus*, which may grow up to 5 feet (1.5 m) long. The madtoms, miniature catfish in the same family, have pectoral spines and poison glands and, in contrast to the blue catfish, grow to only 5 inches (12.5 cm).

North America also has three blind catfish, with atrophied eyes. Two species are found in deep artesian wells and associated ditches near San Antonio, Texas, while the third is found in a well in northern Mexico.

Marine catfish of the family Ariidae are mouth-brooders. The male holds the eggs, which are ¾–2 inches (2–5 cm) in diameter, in his mouth, and when they hatch he continues to shelter the fry in the same way. To do this, he must fast for a month.

The Plotosidae, another family of marine catfish, contains one of the most dangerous fish to inhabit coral reefs. The dorsal and pectoral fins of this species carry spines equipped with poison glands. Merely to brush the skin against these spines can produce painful wounds in humans. Equally dangerous are the parasitic catfish. Some members of this family, the Trichomycteridae, are free-living but many attach themselves to other fish using the spines on the gill covers to hook themselves on, piercing the skin and gorging on the host's blood. Others insinuate themselves into the gill cavities, eating the gills. The tiny candirú, *Vandellia cirrhosa*, of Brazil, has been known to make its way into the urethra of a person entering the water, especially, so it seems, if water is passed. A surgical operation may be necessary to remove the fish.

EUROPEAN CATFISH

CLASS	**Osteichthyes**
ORDER	**Siluriformes**
FAMILY	**Siluridae**
GENUS AND SPECIES	***Silurus glanis***

ALTERNATIVE NAMES
Sheatfish; wels; silurus

WEIGHT
Up to 730 lb. (330 kg)

LENGTH
Usually up to 16½ ft. (5 m)

DISTINCTIVE FEATURES
Long tail tapering to small tail fin; 1 pair of very long barbels on upper jaw, 2 pairs of short barbels on lower jaw; mainly olive green with darker back, pale belly and whitish blotches on sides

DIET
Crustaceans and fish, especially eels, burbot, tench, bream and roach; also ducklings, water voles and amphibians

BREEDING
Age at first breeding: 2–3 years (male), 3–4 years (female); breeding season: May–July; number of eggs: up to 100,000 in large females; hatching period: eggs hatch when ⅓ in. (7 mm) in length

LIFE SPAN
Usually up to 20 years

HABITAT
Associated with overhanging riverbanks and sunken structures, in muddy, stagnant pools, lakes, swamps and backwaters

DISTRIBUTION
Central Europe east to western Russia and northern Iran; also southern Britain

STATUS
Very common

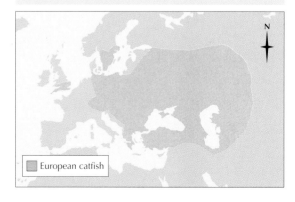

European catfish

CATS

BOTH THE MORPHOLOGY AND THE behavior of cats have been adapted to hunting prey. The special requirements of hunting can be seen in the layout of their skeletons and nervous systems, while the availability of prey strongly influences behavior toward other cats and the ways in which cats communicate. Cats are members of the order Carnivora, and most authorities now recognize 40 species.

Well-adapted hunters

The physiology of a cat's gut has become specialized for eating meat, unlike those of other carnivores which can also feed on plant matter, and cats have evolved over time to become skillful hunters. Hunting demands acute senses to detect and follow prey, superb camouflage and stealthy behavior to avoid detection by the intended target and a high degree of maneuverability to pursue and catch the quarry.

Cats depend primarily on sight and hearing, rather than smell for hunting. Their eyes are adapted to nocturnal and crepuscular (twilight) use; the cheetah, *Acinonyx jubatus*, is alone in the cat family as a regular daytime hunter. The iris is a vertical oval, rather than a circle, as this allows wider opening at night and greater narrowing during the day when less light need be admitted. In addition a reflector, the tapetum lucidum, is located behind the retina of each eye. This reflects any light that has passed through the retina, increasing the cat's ability to see in low light levels.

The short and rounded head common to all cats permits good binocular vision. Leaping onto prey at short distances is a common hunting technique, but one that requires the ability to judge distance very accurately. The flattened face of cats suits binocular vision, as a cat's vision is unimpeded by the large nose common to most other carnivores. The skull shape characteristic of cats is not suited to the large nasal passages needed for a strong sense of smell, and cats do not have as highly developed a sense of smell as other carnivores. The cheetah does have an enlarged nose. This, however, is an adaptation for breathing, not smell, for strong breathing is fundamental to sustain the cheetah's extraordinary running speed. The cheetah's large nasal cavities also help it subdue prey by biting the victim's throat to block the windpipe and cause suffocation. Wide nasal passages enable the cheetah to maintain its breathing while stifling the victim.

The African wildcat, Felis silvestris lybica, *is the ancestor of today's domestic cats. Its coat varies in color according to habitat, being darker in forests and paler in arid areas.*

The cats' senses of hearing and touch are vital complements to their sensitive eyes. Cats can hear sounds up to a maximum of 100 kHz. This is far above the human limit of 20 kHz, and includes the range in which rodents call to one another. However, such high-frequency sounds are of low intensity. To compensate for this, cats also have large pinnae, or ear flaps, which help to pinpoint the source of a sound. The large external pinnae of the sand cat, *Felis margarita*, help it to locate sounds made by its rodent prey in deserts, the hot air of which readily absorbs sound. The serval, *F. serval*, hunts mainly in long grass and is another species with oversized pinnae, which twitch almost constantly to locate the source of faint rustling noises in the grass.

A cat's whiskers have two important functions. At night the wide opening of the eye makes it difficult for the lens to focus on close objects, so the whiskers help the cat to find its

CLASSIFICATION
CLASS
Mammalia
ORDER
Carnivora
FAMILY
Felidae
SUBFAMILY
Acinonychinae, Felinae, Pantherinae
NUMBER OF SPECIES
Genus *Acinonyx*: 1
Genus *Felis*: 27
Genus *Lynx*: 5
Genus *Pardofelis*: 1
Genus *Neofelis*: 1
Genus *Panthera*: 5

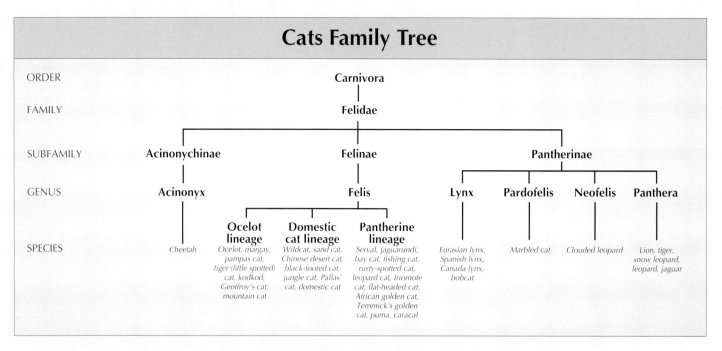

Cats Family Tree

ORDER				Carnivora					
FAMILY				Felidae					
SUBFAMILY	Acinonychinae		Felinae			Pantherinae			
GENUS	Acinonyx			Felis		Lynx	Pardofelis	Neofelis	Panthera
SPECIES	Cheetah	Ocelot lineage	Domestic cat lineage	Pantherine lineage		Eurasian lynx, Spanish lynx, Canada lynx, bobcat	Marbled cat	Clouded leopard	Lion, tiger, snow leopard, leopard, jaguar
		Ocelot, margay, pampas cat, tiger (little spotted) cat, kodkod, Geoffroy's cat, mountain cat	Wildcat, sand cat, Chinese desert cat, black-footed cat, jungle cat, Pallas' cat, domestic cat	Serval, jaguarundi, bay cat, fishing cat, rusty-spotted cat, leopard cat, Iriomote cat, flat-headed cat, African golden cat, Temmick's golden cat, puma, caracal					

way and avoid obstacles in its path. The mystacial (muzzle) whiskers are extended in front of the cat's face when prey has been captured to help locate the best point to bite once the animal has been pinned down.

Camouflage

Hunting cats primarily operate by stalking and ambushing their prey; they must therefore conceal themselves from their quarry. This is achieved in two ways. Cats make use of any available cover, such as vegetation and undulating terrain, and have evolved a variety of disguises in the form of the cryptic patterns on their coats. In most species the coat markings consist of a complex pattern of dark stripes or spots set against a lighter background. Aided by the surrounding vegetation, the effect is to break up the cats' outline. Species native to desert or semidesert regions, such as the sand cat, Chinese desert cat (*F. bieti*), Pallas' cat (*F. manul*) and caracal (*F. caracal*), have plainer, light-colored coats. The coat of the snow leopard, *Panthera unis*, is gray white in color to blend in with high-altitude mountain landscapes dominated by ice, snow and rock.

Locomotion

Cats must be highly agile to catch prey but have evolved a range of skeletal adaptations to suit their particular lifestyle and habitat. Climbing cats such as the margay, *F. wiedii*, of Central America, possess strengthened scapulas (shoulder blades) and ankles that can rotate through 180°. These features allow them to descend trees headfirst, with the hind feet facing up the tree.

All cats are digitigrade, meaning that they walk on their toes rather than on the whole flat foot. This allows an increased stride length, an important feature for those cats that are adapted for running. Stride length is further increased by a flexible backbone

The rare and secretive clouded leopard, Neofelis nebulosa, *lives in the tropical forests of southern and Southeast Asia. Its richly patterned coat mimics the sun-dappled surroundings.*

The tiger is the largest species of cat, and also one of the most endangered. Like the majority of cats, it mainly hunts at night and during the half-light of morning and evening, when its combination of stealth, camouflage and acute senses is most effective.

and a reduction in the size of the collarbone, or clavicle, although this has the concomitant effect of weakening the rib cage of fast-running cats.

When cats are at rest ligaments in their paws hold their claws inside a sheath of skin. This protects the claws from wear and keeps them sharp. Muscles in the paws protract the claws when required for climbing, feeding or fighting. The cheetah uses its claws to provide a sure grip when running at speed; it is also exceptional in that it has no sheath of skin to protect its claws, which therefore usually become blunt. Most cats have a small dew claw on their forefeet; this claw is more developed in the cheetah, which uses it for tripping up prey.

Cats' paws have soft pads that minimize noise when stalking prey. The lower surfaces of the sand cat's paws are also covered with fur to offer protection from extreme high and low desert temperatures. In addition the fur helps to spread the weight of the cat and provide it with some stability when it moves over shifting sands. The Canada lynx, *Lynx canadensis*, is another species with fur-covered feet. In this instance the fur insulates the animal against cold weather and spreads its weight over soft snows.

Territoriality

Most cats lead solitary lives, but communication is necessary for establishing territories, finding mates and controlling social groupings. Smell and sound both play an important role, as these senses permit communication across territorial boundaries without the need for close contact, which might lead to conflict. A wild cat's home range is definable as a group of hunting grounds, drinking and resting places, lookout positions and (for females) sites for dens where cubs or kittens may be reared. These areas are linked together by a series of trails.

A cat's home range varies over time according to a number of factors. Other cats of the same or different species may be tolerated to a greater or lesser extent, although a core area may be reserved and defended rigorously against intruders. Generally, larger cat species have more extensive home ranges than smaller species. The home range of an adult tiger, *P. tigris*, may cover 38 square miles (100 sq km), whereas that of the endangered Iriomote cat, *F. iromotensis*, a native of one of the Ryuku Islands to the south of Japan, typically covers no more than 1 square mile (3 sq km).

The home ranges of male and female cats serve different purposes. A male cat's home range generally overlaps with several female home ranges, though not with the home ranges of other males. The male uses this large area to provide himself with food and to mate with a number of females without suffering competition from rival adult males. This ensures the maximum number of offspring. A female home range, however, is dedicated to denning and rearing young in safety and providing sufficient food for the family.

The sand cat's large ears can be turned into a horizontal, downward-facing position. This may help it detect faint sounds made by prey underground.

Although an adult female allows males into her home range, she will not tolerate intrusions by other females of breeding age, as they would compete for males and food resources.

Making their mark

The most common forms of territorial marking are accomplished by means of pungent odors, and include urine spraying, the depositing of feces and the transfer of secretions from glands on the chin and neck. These signals are often reinforced with visual markings, such as scratching patterns on tree bark and in the earth. However, scent markings outlive scratch marks, and so are more effective.

As a cat's home range varies over time, and the owner uses only certain areas of its home range during a particular period, the cat needs to regularly define areas in which intruders will not be tolerated. Patrolling the trails in its home range, a cat will stop and spray at certain points, creating a visual and olfactory mark. The tiger sprays up to 11 times within half an hour. The retractable penises of male cats enable them to direct their spray at a height level with the noses of other cats, so that there can be little confusion over boundaries. Cats also spray on the ground and scratch over the area to make their mark more noticeable. This behavior is generally limited to flat areas devoid of trees, however.

In contrast to domestic cats, wild cats do not bury their feces, but leave the deposits on logs or other raised surfaces in full view of other cats. However, females tend to bury their feces and refrain from scent marking in their core territory, as alerting intruders to the boundaries of this area also advertises the presence of kittens or cubs to potential predators.

Vocal communication is distinctive in the genus *Panthera*, which contains the tiger, snow leopard, leopard (*P. pardus*), jaguar (*P. onca*) and lion (*P. leo*). These so-called big cats have more flexible cartilage in the hyoid, the structure that connects the voice box to the skull. This enables these cats to roar but they cannot purr, unlike the members of the five other genera of cats. Purring is a quiet call produced when breathing either in or out, and is particularly used for communication between a female and her offspring. Purring is suitable only for communication at short ranges and does not alert predators. The potential predators of small and medium-sized species of cat include hyenas, wolves and other large dogs, bears, birds of prey and larger cats.

Social groupings

Although male cats may roam throughout the territories of several females when mating, true sociality is confined to just 3 of the 40 cat species: the cheetah, domestic cat and lion. The degree of group behavior is always determined by the type and relative abundance of prey.

Sociality is most highly developed in the African lion, which takes large mammalian prey. The lack of good cover in its savanna habitat and the danger of being kicked by powerful prey animals have led to the formation of prides, or groups. Females form the majority of a pride and are responsible for almost all of the hunting, with only a small number of dominant males in the group. There appears to be little hierarchy among the females in a pride, although at a carcass the top male feeds first and the cubs last of all. The size of a pride is related to the supply of prey. In East Africa, where food is plentiful, prides of up to 25 are known; females in this area, in particular the Serengeti, may live with their cubs in groups of 2 to 12 individuals. Females will hunt together for food and nurse and suckle each other's young. They are usually also genetically related to each other. Small bands of younger males that have been excluded from prides also occur and are known as coalitions. However, lions are not invariably sociable: they may live in smaller prides or be solitary.

Coalitions are also found in the cheetah, where cooperation brings its reward in greater body weight due to increased hunting success. Although it is a primarily solitary cat, the tiger is also known to hunt in groups where cover is sparse. This is because individual attacks are more likely to be detected and prove unsuccessful than attacks made in groups.

For particular species see:
- BOBCAT • CARACAL • CHEETAH • CIVIT • GENET
- JAGUAR • JAGUARUNDI • LEOPARD • LION • LYNX
- OCELOT • PALM CIVET • PUMA • SERVAL
- SNOW LEOPARD • TIGER • WILDCAT

CAT SHARK

THE CAT SHARKS BELONG to the family Scyliorhinidae. This is one of the largest families of sharks and contains more than 60 species, including the dogfish. Scyliorhinidae includes many species which are distinguished by their distinctive patterns of stripes, bars and mottling. One species, the South African skaamoog, *Holohalaelurus regani*, is covered with markings that resemble Egyptian hieroglyphics. The patterning is most pronounced in adults. Skaamoogs have broad bodies and their eyes are spaced far apart.

The skaamoog, or skaamhaai (shy eye), and the six species of swell sharks are among the more unusual cat sharks. The former lives off the coast of South Africa, from Port Nolloth to Natal, and has been recorded at depths of 360–1,500 feet (110–460 m). Swell sharks are widely distributed in the seas off the Pacific coast of America from central California to the Gulf of California, central Chile and southern Mexico. The family Scyliorhinidae also contains 13 species of deep-water sharks that belong to the genus *Apristurus*.

There is a related family of false cat sharks, the Pseudotriakidae, which also live in deep water. One species, *Pseudotriakis microdon*, lives at depths of 1,000–5,000 feet (300–1,525 m) in the Atlantic, the other off the coast of Japan. False cat sharks grow up to 10 feet (3 m) long and have an unusually long dorsal fin, but relatively few specimens have ever been seen.

Sharks of the seabed

Cat sharks have two dorsal fins. The upper lobe of the tail is horizontal, not uptilted in the usual shark fashion. There is no nictitating membrane (third eyelid) to the eye. This last feature may be linked with the unusual behavior of the skaamoogs which, when caught, curl their tails over their heads, seemingly in an attempt to hide their eyes. This trait has given rise to their alternative name, shy shark, or shy eye.

Californian swell sharks, *Cephaloscyllium ventriosum*, are relatively well known. They are small, with broad bodies not more than 4 feet (1.2 m) long, and live in shallow seas, especially in beds of kelp. These sharks are light brown in color, with many darker spots and blotches that become gradually more intense toward the tail. They have a large mouth with a set of small, very sharp teeth. Swell sharks prey on small fish, although they probably also take carrion, which often finds its way into lobster traps.

The Californian swell shark is able to inflate to twice its normal size if removed from water. Scientists believe that this trait may be a defense mechanism.

The variegated cat shark, Parascyllium variolatum, *inhabits the deeper waters around Australia. Its ability to camouflage itself makes it almost impossible to see against the seabed.*

Birth of a shark

All members of the cat shark family have similar breeding habits. Fertilization is internal and the female lays eggs enclosed in rectangular, translucent horny cases with a long tendril at each corner. The tendrils become entangled with objects such as the branches of soft corals (sea fans) and are anchored for 8 months (in swell sharks). During this time the embryo develops, feeding on the large supply of yolk contained in the egg. During the latter stages of development the baby shark rotates slowly and spasmodically within the horny case. At the end of 8 months it breaks out at one end of the case.

Swell defense mechanism

Swell sharks are named for their most distinctive feature. When they are hauled out of water, they swell up to twice their normal diameter. To do this the sharks swallow air, which causes the stomach to inflate like a balloon. When a swollen shark is thrown back into the water, it is likely to float until it can discharge the air. The time taken to do this varies. Some will deflate fairly quickly and then swim to the bottom. Others may take as long as 4 days, during which time they float.

Normally sharks do not leave the water voluntarily, and any intake of water will not put the fish at the same disadvantage as a stomach full of air. However, it is difficult to establish how this behavior benefits the shark. Scientists believe that it may be a defense mechanism: by inflating to a diameter of twice its normal size, the shark may well be attempting to deter potential predators from attacking.

CAT SHARKS

CLASS **Chondrichthyes**

ORDER **Carcharhiniformes**

FAMILY **Scyliorhinidae**

GENUS AND SPECIES **More than 60 species, including skaamoog,** *Holohalaelurus regani*, **and swell shark,** *Cephaloscyllium ventriosum*

ALTERNATIVE NAMES
H. regani: shy eye, skaamhaai, izak, lazyshark, leopardshark, tigershark

SIZE
Up to 4 ft. (1.2 m)

DISTINCTIVE FEATURES
First dorsal fin base opposite or behind pelvic fin base; small teeth. *C. ventriosum* distends abdomen by swallowing air or water if brought to surface. *H. regani* has distinctive "hieroglyph" markings on body and fins.

DIET
Wide variety of animals; also carrion

BREEDING
Age at first breeding: when 18–21½ in. (45–55 cm) long (male); number of eggs: 2 egg cases laid at a time; hatching period: usually about 240 days

LIFE SPAN
Not known

HABITAT
C. ventriosum: mainly inshore waters; *H. regani*: coastal waters at depths of 490–525 ft. (150–160 m), juveniles in deeper water than adults

DISTRIBUTION
C. ventriosum: coasts of California, northwestern Mexico and Chile; *H. regani*: coasts of southern and eastern Africa

STATUS
Not known

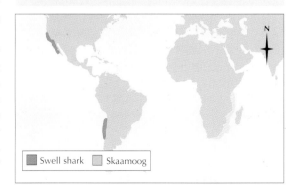

■ Swell shark   ■ Skaamoog

CAVE FISH

THERE ARE 32 KNOWN SPECIES of fish that spend their whole lives in underground caves or artesian wells. Although they belong to unrelated orders and families, they have many characteristics in common. All are small, the largest, the Kentucky blindfish, being 8 inches (20 cm) long, and most are 3½ inches (9 cm) or less. As they have little or no skin pigment, they are mainly pinkish in color, due to the surface blood vessels showing through the scaleless, or nearly scaleless, skin.

Most cave fish are blind. The remnants of eyes are often hidden under the skin, but even in species with visible eyes, these are degenerate, serving only to distinguish light from darkness. The young of cave fish often start with perfect eyes, which degenerate as they grow. In the Mexican cave fish, *Anoptichthys jordani*, the eyes of the adults differ according to where they are living; in isolated caves the fish are totally blind. Those living in caves connected with a surface river have eyes that are almost perfect. All gradations between these two extremes, total blindness or perfect sight, are found in the 32 species according to the situation of the cave and how much light enters it.

Not all sightless fish are cave dwellers. Many deep-sea fish are blind, as are some of those that live among rocks on riverbeds or in swamps. Most cave fish have their nearest relatives in normal freshwater fish, but two species, one in Cuba and the other in Yucatan, are related to deep-sea blind fish. A minnowlike cave fish in Dalmatia is washed to the surface during floods and spawns in the surface waters.

Chemical detectors

Two of the advantages of this underground existence are the absence of predators and the abundance of food. Among the animals living in caves, there are few fish, the majority being numerous invertebrates, especially insects and crustaceans. The only firm information scientists have on the feeding habits of cave fish is that one species lives on insects washed into the cave and another feeds on bat droppings.

Few cave fish have been well studied but it is known that there are marked differences in behavior. The loss of sight is compensated for by the development of the lateral line system, which detects vibrations or changes of pressure in the water, or by an increase in the numbers of

Most cave fish are blind. The abundance of food and lack of predators in their environment, coupled with their sensitivity to movement, means that this is not a major disadvantage.

taste buds. Many cave fish have a more sensitive lateral line, and this is extended on to the head in the form of sensory canals or rows of tiny sense organs, which allow the cave fish to detect obstacles at a distance. In other species, notably in a cave fish from Iraq, the lips and barbels are abundantly sprinkled with taste buds, as many as 260,000 per square inch (40,000 per cm²).

Taste buds are normally found on the tongue. That the cave fish have them on the outer skin means that they are tasting their environment as a means of keeping in touch with their surroundings as well as finding food. External chemoreceptors also allow cave fish to detect the direction from which chemicals are coming.

Finding their way

The Iraqi cave fish moves slowly when it is in still water, by alternating flicks of the tail with glides, at intervals of 1–10 seconds between flicks. In a current it swims with vigorous flicks of the tail. When obstacles are met, the fish shows no evasive action except to turn gently away from them at the last moment.

The Mexican cave fish, by contrast, swims constantly and vigorously, and seems to detect an obstacle from a greater distance than the Iraqi cave fish. It also responds readily to vibrations, such as a tap on the side of the rock basin, whereas the Iraqi cave fish shows no visible response to such stimulae.

Reproduction

Breeding habits also differ. Some cave fish bear live young after a gestation period of 3 months. In other species the female lays eggs and carries them in her gill chambers for 2 months.

The Mexican cave fish has an elaborate courtship in which male and female make exaggerated movements of the mouth and gills. Presumably the turbulence this produces keeps each notified of where the other is. Then the pair abruptly swim side by side and mate, the fertilized eggs sinking to the bottom, where they stick to the rock. All cave fish species have a long breeding season, and most breed year-round.

Blind to save energy

There has been much speculation as to why cave fish should lose the use of their eyes. Living in underground waters, cave fish escape predators and consequently there is reduced competition for the available food. Their food, however, is not always abundant; it varies with the seasons and cave fish must be able to go without food for long periods. It has been suggested, therefore, that by not growing eye tissues, or by allowing eye tissues to degenerate, cave fish economize on energy use.

CAVE FISH

CLASS **Osteichthyes**

ORDER **Cypriniformes (discussed below); Percopsiformes (family Amblyopsidae, including 4 genera with 10 species)**

FAMILY **Cyprinidae**

GENUS AND SPECIES **(1) *Typhologarra widdowsoni*; (2) *Anoptichthys jordani*; (3) *Caecobarbus geertsi***

ALTERNATIVE NAMES
(1) Iraqi cave fish; (2) Mexican cave characin; (3) African cave fish

LENGTH
Up to 3 in. (8 cm)

DISTINCTIVE FEATURES
All 3 species: eyes absent or vestigial, often partly overgrown by skin; body usually pale and flesh-colored, with silver sheen: (2) eyes vestigial (adult), small and functional (young); male slim; female more robust and delicately colored. (3) eyes absent; long, compressed body; sex distinction unknown.

DIET
Any food available

BREEDING
Some species bear live young after 3-month gestation, in others female lays eggs and carries them in gill chambers for 2 months

LIFE SPAN
Not known

HABITAT
Subterranean waters, including pools and streams

DISTRIBUTION
(1) vicinity of Euphrates River, Iraq; (2) near St. Luis Potosi, Mexico; (3) near Thynsville in lower Congo region, central Africa

STATUS
Not known

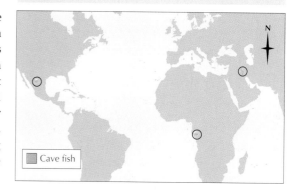

Cave fish

CENTIPEDE

THERE ARE ABOUT 3,000 known species of centipedes but few have individual common names. Scientists, however, have divided the various centipede species into five orders: the Scolopendromorphs (millipedelike forms); Lithobiomorphs (living-under-stone forms); Geophilomorphs (earth-liking forms); Scutigeromorphs (shield-covered forms); and Craterostigmomorphs (species occurring only in Australia). The first two orders include active, elongate and fairly bulky species having respectively fewer than than 24 pairs of legs and 15 pairs of legs. The Geophilomorphs are slender, wormlike centipedes that are adapted to burrowing in the soil. Their legs vary in number from 31 to 177 pairs, so that some of them amply justify the name centipede or "hundred feet."

The Scutigeromorphs are a very distinct order of centipede. The body of these species is short and cigar-shaped, not sinuous as in other species of centipede, and the 15 pairs of legs are very long and slender, enabling the animals to move with great speed and agility. Their respiratory system is significantly more efficient than that of other centipedes, due mainly to the greater oxygen-carrying capacity of their blood.

This feature is directly related to their high rate of activity. The Scutigeromorphs occur in southern Europe and the United States and are widespread throughout the Tropics.

Most centipedes attain a length of only 1–2 inches (2.5–5 cm), but the tropical American *Scolopendra gigantea* may be 1 foot (30 cm) long. The largest centipedes found in the United States are also in the genus *Scolopendra*; individuals may reach 6 inches (15 cm) long. The Asian species, *S. morsitans*, reaches 8 inches (20 cm).

Require a humid environment

The body covering of centipedes is not waterproof and they easily die of desiccation. They are confined, therefore, to humid surroundings and are commonly found in leaf mold and compost heaps, under logs and stones, beneath the bark of trees and in the soil. Centipedes come into the open only at night when the air is moist and cool. One British Geophilomorph, *Strigamia maritima*, lives on the seashore under stones that lie around the high-tide mark.

Centipedes often enter houses and one long-legged species, *Scutigera forceps*, found in dwellings in warm parts of the world, is regarded as

Centipedes are often confused with millipedes but each segment of a millipede has four legs, whereas centipedes have only one pair of legs to each segment.

411

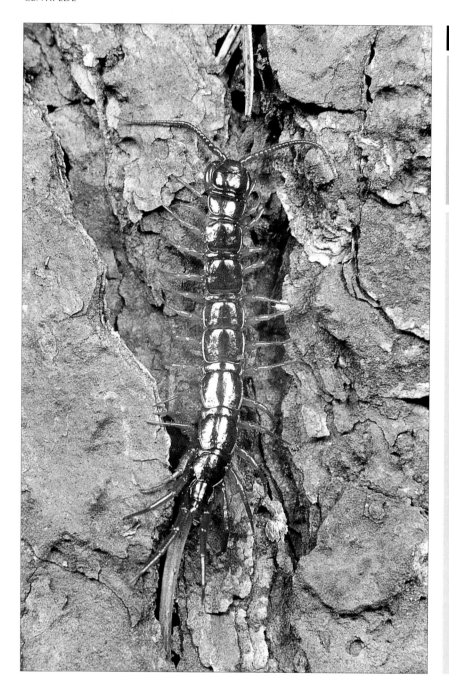

LITHOBIOMORPH CENTIPEDES

PHYLUM	**Arthropoda**
CLASS	**Chilopoda**
ORDER	**Lithobiomorpha**
FAMILY	**Lithobiidae**

GENUS AND SPECIES **Many species, including** *Lithobius forficatus* **(discussed below)**

LENGTH
¾–1¼ in. (1.8–3 cm)

DISTINCTIVE FEATURES
15 body segments, which may appear flattened; 1 pair of legs per segment; chestnut brown in color

DIET
Insects, spiders and other invertebrates

BREEDING
Female lays eggs singly to hatch unattended; young hatch with 7 pairs of legs and develop the further 8 pairs of adults through successive molts

LIFE SPAN
Up to 5–6 years

HABITAT
Usually woodland, grassland and moorland; also on sea shore above high-tidemark and up to 1,500 ft. (460 m) in mountains

DISTRIBUTION
Temperate regions of both Old World and New World

STATUS
Many species abundant

The tiny species **Lithobius forficatus** *hunts small worms, spiders and flies.*

beneficial to the household as it preys on insects. In tropical Asia the large and venomous *Scolopendra morsitans* is often found in houses, but its poisonous bite means that it is by no means a welcome visitor. Many caves in hot countries harbor centipedes, especially the long-legged Scutigeromorphs. The cool of the caves provides the creatures with a refuge from the harmful drying effects of the sun.

Active hunters

Centipedes are active predators, hunting insects, spiders, worms and other small prey. Some Geophilomorphs eat plant material as well and have occasionally occurred in such large numbers as to be considered a pest. The common species *Lithobius forficatus* readily eats flies in

captivity and a large tropical Scolopendra which was kept for over a year in London Zoo fed mainly on small mice. In the wild state these large centipedes prey on large insects such as locusts and cockroaches and also on geckos and other nocturnal lizards. The marine centipede *Strigamia maritima* has been observed to feed on barnacles along the shore.

Venomous legs

Small centipedes are eaten by birds when exposed by spade or plow but are protected from animals of their own size by their poisonous bite, which paralyzes and kills prey. The bite is delivered by the front pair of legs, known as maxillipeds. These have become modified over time to serve as poison jaws; the legs are hollow and

connected to poison glands. The bite of the large tropical Scolopendras species is excessively painful and occasionally dangerous. Many Scolopendras are brightly colored; their striking appearance acts as a warning to other animals that they are venomous. However, very few human fatalities have been recorded as a result of their bite, and the jaws of smaller centipedes are not large enough to puncture human skin.

Breeding strategies

The sexes of centipedes are very similar in appearance and can usually be distinguished only by microscopic examination. All species of centipede lay eggs. The Lithobiomorphs and Scutigeromorphs lay their eggs singly in the soil and leave them to develop unattended. When their young hatch they have only seven pairs of legs, reaching the full number in the course of development. Geophilomorphs and Scolopendromorphs lay 15 to 60 eggs in nests.

The big species of Scolopendra brood and guard their eggs and young, fighting any enemy that attacks, and also protect them from fungal infection by licking them. If adult Scolopendras become seriously disturbed, they often eat their eggs and young, or may desert the nest; unattended, the eggs usually go moldy and die.

Centipedes are long-lived and even the diminutive *Lithobius forficatus* may live for 5 or 6 years. The large tropical species probably take at least 4 years to reach full size and may live for considerably longer than this.

See how they run

Centipede locomotion was not understood until the advent of high-speed photography. In most centipedes the legs move rhythmically in waves, which alternate on either side of the body, so that in any part, at a given moment, the feet will be bunched on one side and spread on the opposite side. When the legs bunch, the tips often cross. In the wormlike Geophilomorph centipedes this pattern of movement is not seen, however; each leg seems to move on its own, picking its way independently of the others. Geophilomorphs are soil-dwelling and much of their movement consists of burrowing rather than moving at speed over land. An independent leg action is more suitable for pushing the body along in soil.

Scutigeromorphs are hyperactive centipedes with very long legs suited to rapid movement.

CHAFFINCH

Chaffinches build their cup-shaped nests in secluded spots, usually low down in a hedge, bush or tree.

THE CHAFFINCH IS A familiar bird throughout much of Europe. It is easily distinguished in flight by a broad white patch on the wings and white streaks on the outer tail feathers. The male has a slate blue crown and nape, chestnut back and cheeks and pinkish brown underparts. The wings and tail are brownish black, with prominent white wing bars. The female's plumage is browner and lacks the slate blue on the head.

In Europe, the continental chaffinch occurs north to the northernmost tree limit in Scandinavia and east to Tomsk in central Russia. Its cheeks and underparts are a purer pink than those found in Britain. British and continental chaffinches mix from mid-September to mid-November, when large flocks of chaffinches migrate from the continent and spread across Britain. Continental chaffinches are also found in northwest Africa, the Canary Islands, the Azores and Madeira. Northernmost breeders migrate southward for the winter; most birds that breed in central and southern Europe are resident.

Feeding flocks

Outside the breeding season chaffinches become gregarious, forming flocks of several dozen. They fly with the undulating flight typical of finches. The wings are flapped rapidly for a few beats, then closed and the bird loses height, climbing again during the next series of wingbeats.

Many birds are more sociable outside the breeding season, particularly in hard weather. But even in mild weather chaffinches form flocks, feeding together in fields and woods and roosting in shrubberies or hedgerows. Chaffinch flocks often join other species, and flocks several hundred strong feed together in the fields. Along with the chaffinches there may be greenfinches, linnets, sparrows, buntings and several species of tit. The drab female chaffinches are less easily detected than the brighter males among a crowd of sparrows or other small birds in a mixed flock. As a result, it was once thought that the hens migrated south, leaving the cock chaffinches behind, which earned the cocks the nickname "bachelor birds."

Seed-eating bill

Chaffinches feed on a variety of seeds, fruit and invertebrates, including insects, spiders and earthworms. More plant than animal food is eaten. At times they take fruit, such as blackberries, apples and currants. Chaffinches also eat blossoms and buds, but never to the extent of being a serious pest, unlike bullfinches, which are closely related. The upper half of a finch's bill has a groove down each side into which the sharp edge of the lower half fits, forming a highly effective nutcracker. The conical bill and large jaw muscles are used for cracking seeds. The chaffinch turns the split seed between the edges of its bill and uses its tongue to peel the husk.

Woodland breeder

In February the winter flocks break up and chaffinches move from open country into woodland. Here the males establish their territories by singing from favored perches and chasing away intruding males. A female is greeted by an invitation posture, the male approaching her in a lopsided crouch with one leg bent. He then lures her into his territory by singing and displaying; if successful, mating takes place.

The female chooses a well hidden site for the nest and builds it unassisted, using grasses, roots and moss. She decorates the outside with pieces of lichen fastened with spiderweb. Nest-building may take 3–18 days and a female chaffinch makes, on average, 1,300 collecting trips. Clutches of four or five eggs, sometimes three to six, are laid in April, May or June, and the female incubates them alone. The chicks hatch after 11–14 days and the female continues to brood and feed them by herself. Later the male helps her, but he supplies only 15 percent of the food.

CHAFFINCH

CLASS	**Aves**
ORDER	**Passeriformes**
FAMILY	**Fringillidae**
GENUS AND SPECIES	***Fringilla coelebs***

WEIGHT
**Male: ¾–1 oz. (20–29 g);
female: ⅔–1 oz. (18–27 g)**

LENGTH
**Head to tail: 5¾ in. (15 cm);
wingspan: 9⅔–11¼ in. (25–29 cm)**

DISTINCTIVE FEATURES
**Male: slate blue nape and crown, pinkish
buff underparts, prominent white wing bars
on darker upperparts; female: much browner**

DIET
Plant material, especially seeds; insects

BREEDING
**Age at first breeding: 1 year; breeding
season: March–July; number of eggs:
usually 4 or 5; incubation period: usually
12–13 days; fledging period: 13–14 days;
breeding interval: 1 year**

LIFE SPAN
Probably up to 8 years

HABITAT
Woodland, parks, gardens, steppe borders

DISTRIBUTION
**Most of Europe, east to Central Asia; winter
range includes North Africa and Middle East**

STATUS
Very common

Chaffinch

The nestlings are fed only insects, because dry seeds are impossible for them to digest. After 2 weeks the young leave the nest and are led to dense cover in the territory. Here they are fed for about 3 weeks; then they become more active and learn to fend for themselves.

Alarm calls

Hawks and owls prey on adult chaffinches, while ermine, cats, magpies and jays rob the nests. The behavior of chaffinches when they see a predator depends on their position. A hawk flying overhead causes them to bolt for cover, where they freeze. However, a perched owl attracts a crowd of chaffinches and other small birds, which gather around it calling loudly. The chaffinches utter a *chink* call when mobbing an owl. This alerts other birds. When hiding from a hawk, the call is a high, thin *seet* that makes other chaffinches fly to cover. Unlike the *chink* call, it is very difficult to locate the source of the *seet* call is. The caller's hiding place is thus not disclosed, but it is enough to warn the other birds that there is a hawk about.

Learning by imitation

The function of a male chaffinch's song is to keep other males out of its territory and to attract females. Each song differs slightly, and can be recognized by other chaffinches.

Many birds can imitate noises made by other birds or by humans and other animals, but chaffinches pick up sounds only from other chaffinches. From early in its life a chaffinch hears the song of its father and other neighboring males. Later, the chaffinch learns the fine detail of the song when it has to sing in competition with neighboring territory-holders.

The song is not completely learned, however. If chaffinches are reared in isolation from a very early age, they are still able to sing, although their song is simpler and quieter than that of a bird reared by its natural parents. This experiment shows that the song is derived from a mixture of experience and instinct. There is a simple, instinctive framework on which the intricacies of the full song are built as the bird hears and practices them.

Male chaffinches have cheerful musical songs with strong regional dialects.

CHALCID WASP

Most chalcid wasps are parasites and often lay their eggs inside the eggs of other insects. Some of them are among the smallest known insects.

CHALCID WASPS ARE MINUTE insects which belong to the order Hymenoptera. They are related to the bees, wasps and ants. Approximately 18,600 species have been described and it is thought that there are others yet to be discovered. Some chalcid wasps, the fairy flies of the family Mymaridae, are among the smallest known insects; one species, *Alaptus magnanimus*, is only 0.12 millimeters long. Other species grow to over 25 millimeters in length.

Parasitic lifestyle

Many chalcid wasps are parasites and hyperparasites of other insects. The term hyperparasite means "parasite of a parasite." Thus a parasitic ichneumon wasp may lay its eggs in a caterpillar and its larvae may, in turn, be parasitized by a smaller wasp, such as a chalcid. Cases are even known of hyperparasites themselves being parasitized. Many chalcid wasps lay their eggs inside the eggs of moths and butterflies. This gives an indication of how tiny some chalcid wasp eggs are. From a single butterfly or moth egg 20 or more chalcid wasps may emerge, having hatched as larvae and pupated into adults within the host egg. The host dies during this process.

Chalcid wasps are of significant economic importance as a means of controlling insect pests. One species, *Pteromalus puparum*, lays its eggs in the newly formed pupae of white butterflies, destroying great numbers of them. In 1929 the small white butterfly, *Pieris rapae*, one of the commonly known cabbage white butterflies, was accidentally introduced into New Zealand and by 1936 was established as a serious pest. In New Zealand the small white butterfly enjoyed a great advantage in that the parasites which had controlled its numbers in Europe were absent. As a remedy, some 500 parasitized pupae of the butterfly were sent to New Zealand from England and more than 12,000 chalcid wasps emerged from these pupae. Most of these adults were released and they, in turn, laid their eggs in the pupae of cabbage whites. Very soon it was discovered that almost 90 percent of pupae collected in the wild had become parasitized, and the *Pieris rapae* has been well under control in New Zealand since that time.

Another example of a pest that has been brought under control by the introduction of chalcid wasps is provided by the whitefly, *Trialeurodes vaporariorum*. This fly has spread to cooler climates, where it feeds on crops grown in greenhouses, such as tomatoes. It can be controlled by fumigation or by biocontrol using the chalcid wasp *Encarsia formosa*. These parasites do not eliminate the whitefly but help to keep them at a manageable level so that damage caused is minimal. Not all chalcid wasps are parasites of insects, however. Some form galls in plants and others bore into seeds and may prove harmful on this account.

Adult chalcid wasps can take only liquid food and probably confine themselves to sucking water, nectar and honeydew. The diet of the larvae depends on their habits; parasitic species feed on the tissues of their insect host, others on the tissues and sap of plants.

Young without fertilization

Many of the chalcid wasps reproduce by a method known as parthenogenesis. This means the females lay eggs that hatch without ever being fertilized. In such species the males are usually rare and sometimes unknown to science.

A more remarkable method of reproduction seen in some chalcid wasps is that known as polyembryony. A single egg is laid by the parasite and at an early stage in its development it divides into a number of separate cell masses, each of which develops into a larva. The individuals formed in this way are all of the same sex and genetically identical. They are formed in the same way as human identical

CHALCID WASP

PHYLUM	**Arthropoda**
CLASS	**Insecta**
ORDER	**Hymenoptera**
SUPERFAMILY	**Chalcidoidea**
FAMILY	**Aphelinidae**
GENUS AND SPECIES	***Encarsia formosa***

LENGTH
Female: less than 1 mm

DISTINCTIVE FEATURES
Tiny size. Female: black head and thorax; yellow abdomen. Male: darker in color.

DIET
Adult: mainly flower nectar and honeydew. Larva: juices of whitefly hosts.

BREEDING
Number of eggs: 60 to 70, laid inside bodies of whitefly hosts; egg to adult: about 25 days

LIFE SPAN
Not known

HABITAT
Areas with whitefly, such as tropical regions and greenhouses in cooler climates

DISTRIBUTION
Almost worldwide

STATUS
Often abundant

twins, but whereas human twins are relatively rare, identical broods from one egg are normal for many chalcid wasps.

Parasitic chalcid wasps usually lay their eggs in the eggs of particular host insects. They are known to attack hosts from 13 insect orders, as well as the egg sacs of spiders, ticks and mites, the cocoons of pseudoscorpions and the galls of certain nematode worms. Some species lay in host eggs in or on the bark of trees; others parasitize the eggs of sawflies and other insects found in galls. Some fairy flies enter water and lay their eggs in those of dragonflies and backswimmers or in the larvae of caddis flies. Only one chalcid lays in any one host egg.

A chalcid wasp can develop from an egg to an adult less than 1 month, although the rate of development depends on temperature. One of the shortest life cycles of any chalcid wasp occurs in the species *Euplectrus comstockii*, which develops from egg to adult in 7 days.

The wasp and the fig

The finest edible figs in the Mediterranean region are those of the Smyrna fig tree. However, the tree bears only female flowers and will not form fruit without pollination. If fruit is to develop the wild fig, or caprifig, the fruit of which is not edible to humans, must be grown among the Smyrna fig trees; bunches of caprifigs can be gathered and hung among the branches of the edible fig trees to effect the same result. This technique was known to the early cultivators of classical times, although they could not explain how it worked.

Scientists now know that certain chalcid wasps in the family Agaontidae, which are also known as fig insects, live in galls on the male flowers of wild figs. The male wasps are wingless and can only crawl from one gall to another in search of females, with which they mate. After mating, the winged females search for fig flowers in which to lay their eggs, and during this search they often enter the flowers of the Smyrna figs. The females do not lay eggs in these flowers but, having emerged from male wild fig flowers, carry on their bodies pollen that fertilizes the Smyrna fig flowers so that they develop and form fruit. Since the wasps can breed only in the wild figs, it is necessary to cultivate these, and to ensure that they are infested by fig wasps, in order to obtain fruit from Smyrna figs. These fig trees are not pollinated by the wind or by any other insects. When fig growing was established in California, fig wasps were imported from the Mediterranean region to ensure successful pollination.

Due to their parasitic lifestyle, chalcid wasps may be constructively introduced into a new environment to control certain pests.

CHAMELEON

Adult chameleons generally take little interest in their young. Some females may even eat their offspring if no alternative food source is available.

THE CHAMELEONS ARE A family of lizards renowned for several unusual features. The body is flattened from side to side and so is relatively high in proportion to its length. The tail in most species is prehensile, often held in a tight coil and can be wrapped around a twig for extra grip. The toes of each foot are joined, three on the inside of the front feet and three on the outside of the hind feet, and can give a tenacious grip on perches.

A chameleon is best-known for three things: its ability to change body color, its eyes set in "turrets" that can move independently of each other and its extensible tongue, which can shoot quickly out to a length greater than the chameleon's head and body. Some species have rows of tubercles down the back, a "helmet" or casque like the flap-necked chameleon, *Chamaeleo dilepis*, or horns like Jackson's chameleon, *C. jacksoni*. A few species grow to 2 feet (61 cm) long, while dwarf species measure less than 4 inches (10 cm).

Most of the 80 species of chameleon live in sub-Saharan Africa, including Madagascar. The common chameleon, *C. chamaeleon*, ranges from the Middle East along the North African coast to southern Spain. Two other species live at the southern end of the Arabian peninsula and a fourth is found in India and Sri Lanka.

Slow-motion movers

Chameleons live mainly in forests and seem to spend most of their time virtually rooted to the spot, the only movement being of the eyes, each independently sweeping from side to side searching for food and danger. When chameleons move, they gradually creep along a twig. A forefoot is released on one side and the hind foot on the other, and both are slowly moved forward to renew their grip on the twig; then, equally stealthily, the other two feet advance. Although most chameleons keep to the trees as much as possible, the stump-tailed chameleons of the genus *Brookesia* can often be found on the ground foraging among leaf litter.

Chameleons periodically shed their skins. Before the old skin comes off, it separates from the new skin under it, leaving an air-filled gap that gives the chameleon a pale, translucent appearance. Then the old skin splits, initially just behind the head, and pieces of it flake off, exposing the brilliant new skin. Even the skin on the eyelids is replaced.

Fast food

Chameleons have a diet similar to that of most small reptiles, consisting of insects and other small invertebrates, but the larger species also catch small birds, lizards and mammals. The method of capture, however, is shared only by the frogs and toads. Chameleons capture their prey by shooting out their long tongue, trapping the victim on the tip and rapidly carrying it back to the mouth. The whole action is so rapid that high-speed photography is needed to show the mechanism at work. Using this technique, it has been found that a tongue 5½ inches (14 cm) in

CHAMOIS

THE CHAMOIS IS THE name given to a species of goat-antelope. It belongs to a group of goatlike mountain animals that includes the gorals and serows of Asia. The chamois is sturdy, 35–50 inches (90–130 cm) long, 30–32 inches (76–81 cm) at the shoulder, and weighs up to 110 pounds (50 kg). The does are smaller than the bucks and both have horns, up to 10 inches (25 cm) long, that rise vertically from the head, then curve back and down to end in sharp points. The coat consists of long hair with a thick underfur, tawny in summer, dark brown to black in winter. There is a patch of white on the throat, some white on the face and a dark line along the middle of the back.

Chamois live mainly in alpine forests at about the level of the tree line and only occasionally venture above this height, though some will range as high as the snow line in summer. At the approach of winter there is a downward movement among the herds, some descending as low as 2,400 feet (730 m). None goes higher than 4,500–6,000 feet (1,370–1,830 m), whereas ibex, a species of mountain sheep found in Eurasia, will move up to 7,000–8,500 feet (2,130–2,600 m).

The chamois ranges through many of the high mountain ranges of southern Europe into Asia Minor and the Caucasus. In Europe it is found in the Cantabrian Mountains, Pyrenees, Alps, Apennines, Sudeten, Tatra Mountains, Carpathians and Balkans. Several subspecies of chamois have evolved, each particular to a mountain region. The does and young live in groups of 15 to 30, which merge to form herds several hundred strong for the rut. The bucks are solitary except during the rutting season.

Spring-heeled

Chamois hooves are made of pliable horn, the inner part of each hoof acting like the welt nails on a climbing boot. Cup-shaped depressions on the soles of the hooves give a firm hold on rocks, and the points of the hooves provide grip and help to prevent slipping on any surface but icy slopes. The hoof also moves up and down like a spring or a shock absorber. This allows the leg to take the strain of a heavy jump.

A chamois is able to leap upward to a height of 6⅔ feet (2 m) or make a long jump of 20 feet (6 m), but even more striking is its ability to

Chamois inhabit high mountains, spending most of their time at the tree line. However, they venture almost to the peaks in summer and may be forced down to lower slopes by very heavy snows.

The chamois' pliable hooves are specially adapted to give the animal a sure grip on all but the iciest slopes.

navigate treacherous rocky crags or to dash across an almost sheer rock face. Apart from their leaping powers, chamois are noted for their remarkable sight and hearing, which makes them a difficult quarry to catch.

Chamois have a goatlike appearance and seem to have the digestive powers of a goat but are in fact bovids. They are grazers and browsers, taking grasses and other sparse mountain herbage, lichens, leaves and pine needles.

Male rivalry

The rut begins in mid-October and lasts until December. The bucks chase each other, erecting the hair along the midline of the back and bleating with a deep rumbling note. The scent glands on the top of the head in both sexes are at their peak of activity at this time. The rivalry between males leads at times to one of the adversaries having his belly punctured by the other's horns, resulting in a fatal peritonitis. After a gestation period of 170 days one to three kids are born from mid-May to mid-June, by which time the herds have once again split up into small groups. The kids are fit enough to follow the does almost from birth. Females mature at 18–24 months.

Potential enemies are any wolves or lynx that reach these heights. The kids are also at risk from foxes and eagles. Humans, however, pose the biggest threat to the chamois, and four subspecies are now endangered or vulnerable.

CHAMOIS

CLASS	**Mammalia**
ORDER	**Artiodactyla**
FAMILY	**Bovidae**
GENUS AND SPECIES	***Rupicapra rupicapra***

WEIGHT
53–110 lb. (24–50 kg)

LENGTH
Head and body: 35–50 in. (90–130 cm); shoulder height: 30–32 in. (75–81 cm); tail: 1¼–1⅗ in. (3–4 cm)

DISTINCTIVE FEATURES
Coarse winter fur; black stripe from each eye to nose; slender, curved black horns

DIET
Mainly grasses, herbs, lichens, moss, leaves and pine needles

BREEDING
Age at first breeding: 8–9 years (male), probably 18–24 months (female); breeding season: October–December; number of young: usually 1; gestation period: 170 days; breeding interval: probably 1 year

LIFE SPAN
Up to 22 years in captivity

HABITAT
Alpine meadows, wooded slopes and crags in high mountains, usually near to tree line

DISTRIBUTION
High mountains of southern, central and eastern Europe, including the Pyrenees, Alps and Tatra Mountains, and east through Caucasus to Asia Minor

STATUS
Generally uncommon, and rare in some areas; *R. r. cartusiana*: critically endangered; *R. r. pyrenaica* and *R. r. tatrica*: endangered; *R. r. rupicapra*: vulnerable

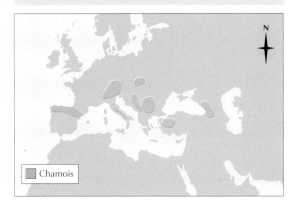

Chamois

CHAPARRAL

CHAPARRAL IS A BIOME primarily composed of small-leaved evergreen shrubs, bushes and dwarf trees that form a dense layer of vegetation 3–13 feet (1–4 m) high. The term chaparral comes from the Spanish *chaparro*, which denotes a covering of shrublike evergreen oaks common to the Mediterranean region of southern Europe. Following the Spanish conquest of the Americas, the term was also applied to the dense brushlands of California, as the vegetation growing there appeared to be very similar to that in some areas of Spain.

Today the word chaparral is often used to refer to the vegetation covering the foothills of the Rocky Mountains. However, in this context the term is not strictly accurate. Rocky Mountain brushlands are winter deciduous (they drop their leaves in the fall) and summer active (they flourish during the summer, when conditions are favorable for growth). The evergreen California chaparral, however, is winter active and summer dormant.

In summer, chaparral plants slow down their metabolic rate and temporarily cease growth in the dry, hot conditions, thus reducing their water intake. In this respect chaparral plants mirror the behavior of desert plants, which also have to economize on water use to survive in arid conditions.

The true chaparral biome occurs in extensive but broken stretches on hillsides and mountain slopes from southern Oregon through to the coastal ranges of California and the foothills of the Sierra Nevada. Chaparral is most common in southern California and the northern third of Baja California in Mexico. It is also found in northwestern Arizona. In California alone, chaparral occupies 7.5 million acres (3 million ha). The area is commercially unimportant but vegetation plays an important ecological role in the Californian chaparral, offering protection for watersheds in regions with steep, easily eroded slopes. Without this vegetation, streams, rivers and rainfall would remove much of the soil.

Climate and conditions

The chaparral is a harsh environment characterized by hot, dry summers and mild, wet winters. In summer, temperatures can soar to as much as 104° F (40° C), and annual precipitation is typically 10–25 inches (25–64 cm), 85 percent of which falls between late October and mid-April.

Shown here are the Santa Catalina Mountains, southern Arizona. Cacti and small, tough-leafed shrubs are typical of chaparral vegetation.

Many types of chaparral are recognized by their dominant shrub species. These types include chamise chaparral (chamisal), manzanita chaparral and desert chaparral. Montane chaparral is adapted to colder climates and extends through coniferous forest zones, such as those in the Laguna Mountains of southern California, up to an altitude of 10,000 feet (3,000 m). Sometimes the word chaparral is used instead of a more general term, Mediterranean vegetation, to describe biomes in other continents, such as those around the Mediterranean Sea (mattoral), African Cape (fynbos), southwestern Australia (mallee) and Chile and Peru (espinal).

Chaparral plants

Several hundred species of plants characterize chaparral. Most evergreen shrub species belong to families such as the sages, oaks, heaths and buckthorns. Chamise (*Adenostoma fasciculatum*), scrub oak (*Quercus dumosa*), sumac (*Rhus* spp.) and manzanita (*Arctostaphylos* spp.) are the dominant plants in the North American chaparral. Common succulents include the prickly pear cactus (*Opuntia* spp.) that occurs in the more arid chaparral communities.

The intense heat of summer has led to the evolution of special adaptations to life in the chaparral. Plants grow small, thick leaves and are able to regenerate quickly after forest fires. The leaves are the most distinctive features of California chaparral shrubs. They are usually rigid, waxy and evergreen, and are commonly referred to as sclerophyllous ("hard leaves"). The leaves' small size and waxy cuticle help to prevent water

This roadrunner is building its nest in a cholla cactus. A native of chaparral country, the roadrunner feeds on insects, lizards and other small animals.

CHAPARRAL

ALTERNATIVE NAMES
Various names used for similar biomes in other continents, including mattoral, fynbos, mallee and espinal

CLIMATE
Mediterranean type: hot, dry summers and mild, wet winters. Annual rainfall: 10–25 in. (25–64 cm); maximum temperature: 104° F (40° C); minimum temperature: 50° F (10° C)

VEGETATION
Mainly dwarf evergreen oaks and woody shrubs; cacti and other succulents

LOCATION
California; northwestern Arizona; northern Baja California; the Mediterranean; central Chile; southwestern Australia; Cape of Africa

STATUS
Vulnerable to bushfires

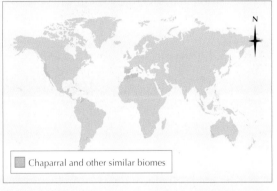

Chaparral and other similar biomes

evaporation from the plants, while hard cells in the leaves maintain their rigidity. Being evergreen and rigid, chaparral plants are in a position to take advantage of rainwater as soon as it arrives and need not use it to inflate wilted leaves, unlike deciduous plants.

Fires and regeneration

Mature areas of chaparral, particularly those over 70 years old, often accumulate a high percentage of dead wood and have little new growth. This condition is attributed to the buildup in the soil of plant substances that inhibit the germination of seeds and the activity of soil microorganisms.

Chaparral is one of the most fire-susceptible environments in the world. This is due to a combination of factors, such as the extremely dry conditions that arise in chaparral regions at the end of the summer, the dense thickets of vegetation that occur there and the high concentration of volatile compounds in the soil.

Chaparral soils are most often rocky and low in nutrients and do not retain water well. On steep mountain slopes the soil tends to erode in summer, especially after the slopes have been scorched by the brushfires that rage during the summer months. However, brushfires can be useful to chaparral regions by removing dead vegetation and converting it into soil-enriching ash. By burning away plant cover, fires also enable sunlight to reach the earth and encourage new growth. Bacteria in the soil are also much more plentiful after a brushfire.

Fires are frequent during the late summer, but they are necessary for the germination of many fire-resistant seeds. These seeds have a hard shell that enables them to survive brushfires, which can create temperatures as high as 1,300° F (700° C). The seeds remain dormant until temperatures drop low enough and conditions become moist enough for new germination to take place. The regrowth of vegetation on these fire-damaged slopes is a process known as secondary succession.

California chaparral does not follow the pattern of classic plant succession. Instead of a series of plant species following other species over time, every plant that comprises the mature chaparral vegetation is already present the first

The coyote is the largest predator living in North American chaparral. It feeds mainly on rodents, insects and carrion.

wet season after a fire. In the first few years, many "fire" herbs are present, but gradually the shrubby seedlings and sprouts grow over these herbs. It takes only 10–20 years following a serious fire for the vegetation to develop into mature chaparral. If too many fires occur, however, the fragile chaparral biome may become grassland.

Chaparral animals

In common with the members of the chaparral plant communities, many of the animal species that inhabit chaparral are adapted for survival in prolonged summer droughts and during hot summer days. Most chaparral animals avoid the intense heat of the day and are active during the cooler hours of night, late afternoon or early morning. Some species spend only part of the year in the chaparral biome, moving elsewhere when conditions become unfavorable. For example, deer and birds usually inhabit the biome only during the wet season, when the growth of new plants provides a plentiful food source. They migrate to higher altitudes as the summer season arrives and food becomes scarce.

Birds that favor chaparral conditions include the roadrunner, *Geococcyx californianus*, a ground-feeder that preys on insects, scorpions, lizards and other small animals; the scrub jay, *Aphelocoma coerulescens*, a noisy species that eats insects and acorns; and the California thrasher, *Toxostoma redivivum*, which uses its strongly decurved bill to sift through sand and dead leaves in search of hidden invertebrates. The wrentit, *Chamaea fasciata*, is virtually confined to chaparral, where it is a common resident.

Rabbits, rodents and deer are the main mammals found in chaparral. The mule deer, *Odocoileus hemonius*, flits through the relatively open woodland to search for food and to avoid the coyote, *Canis latrans*. The gray fox, *Urocyon cinereoargenteus*, emerges from its underground burrows at nightfall to hunt for small rodents.

Alligator lizards in the genus *Gerrhonotus* and several species of rattlesnakes, including those in the genera *Crotalus* and *Sistrurus*, are found throughout the Californian chaparral. They feed on rodents such as the California mouse, *Peromyuscus californicus*, and on small mammals such as the brush rabbit, *Sylvilagus bachmani*. Chaparral scrub also protects a range of smaller species, especially insects and other invertebrates, reptiles and rodents, from attack by aerial predators. These include the red-tailed hawk, *Buteo jamaicensis*, and the Harris hawk, *Parabuteo unicinctus*.

The western diamond-back rattlesnake, Crotalus atrox, is nocturnal. In spring, however, it forages during the late afternoon, because nights are too cold for it to hunt.

CHEETAH

T HE CHEETAH IS DISTINGUISHABLE from other big cats by its slim build. The legs of a cheetah are disproportionately long when compared with other cats, and the head is small. Cheetahs grow to 3½–5 feet (1.1–1.5 m) long and stand up to 3 feet (90 cm) high; they may weigh more than 154 pounds (70 kg). The ground color of the cheetah's coat is tawny to light gray with white underparts. Most of the body is covered with closely spaced black spots, which merge into black rings on the tail. On each side of the face there is a distinctive black "tear stripe" running from eye to mouth.

In 1927 a second species of cheetah was described from Zimbabwe. It was called the king cheetah, *Acinonyx rex*, and had black stripes replacing some of the spots, longitudinal on the back and tail and diagonal on the flanks. However, the claimed species was based on only a few specimens and in retrospect was no more than a local aberration rather than a true species. Although similar mutant cheetahs have been observed over time, including white cheetahs with bluish spots and individuals with marbled coats, all cheetahs are very similar genetically and there are not currently thought to be any identifiable cheetah subspecies.

The world's fastest animal

The cheetah is traditionally thought of as the fastest animal on land and has a body adapted for speed. Its long legs and flexible spine allow for a lengthy stride, which provides extra speed. Exposed claws act like running spikes to give the cheetah stability at high speeds, while the long tail helps to counterbalance the animal when turning quickly. Hard footpads enable the cheetah to run rapidly over hard ground, while its highly developed dew claws help the cat bring down its prey.

Cheetahs have been estimated to run at speeds of up to 70 miles per hour (115 km/h). However, such high-speed sprints can only be maintained for a few hundred meters. Generally a cheetah will stalk prey to within 160 feet (50 m) and only then launch an attack.

Hunting on the run

The cheetah is a diurnal hunter, active during the early morning and late afternoon. In Africa its main prey are impala, springbok, kob and Thomson's and Grant's gazelles. Small animals such as hares probably also form a substantial part of its diet. Two or more male cheetahs are able to attack larger animals such as wildebeest

Cheetahs may reach speeds of 70 mph (115 km/h) during a chase, but only for a limited distance. Most cheetah hunts end in failure.

CHEETAH

CLASS	**Mammalia**
ORDER	**Carnivora**
FAMILY	**Felidae**
GENUS AND SPECIES	***Acinonyx jubatus***

WEIGHT
Usually 80–145 lb. (36–66 kg)

LENGTH
Head and body: 3½–5 ft. (1.1–1.5m); shoulder height: 28–35 in. (70–90 cm)

DISTINCTIVE FEATURES
Slim build; long legs; small head; black stripe from eye to mouth; small round spots on coat; long, spotted tail, striped at tip

DIET
Small antelope, calves of large antelope, hares and other small mammals, gamebirds

BREEDING
Age at first breeding: 2 years; gestation period: 90–95 days; number of young: usually 3 to 5; breeding interval: 1–2 years

LIFE SPAN
Up to 16 years in captivity

HABITAT
Mainly open grassland and savanna scrub; also semidesert and forest fringes

DISTRIBUTION
Sub-Saharan Africa; probably also Iran

STATUS
Vulnerable; probably endangered in northwestern Africa and Iran

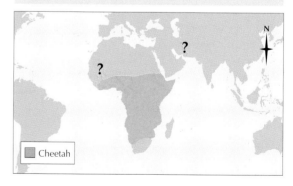

Cheetah

Female cheetahs are solitary and do not maintain strict territories. However, males often form groups of up to four individuals. These groups are known as coalitions, and their members do not tolerate rivals in their area.

and zebra. Young ostriches and gamebirds, including bustards, button quail and guinea fowl, are also eaten.

Unlike other big cats, which tend to lie in wait and pounce with a single leap or a short rush, cheetahs stalk their victims and then race after them. In a short sprint they can easily overtake their prey, but if the latter gets a good start, the cheetah will drop out of the chase, exhausted from its violent burst of energy. Few chases last more than a minute.

Having caught up with its prey, the cheetah brings it to a halt and dispatches it by seizing its throat. The cheetah's strong jaws can clamp another animal's neck and hold tight for up to 20 minutes until the prey suffocates. The cheetah's nasal cavities, larger than usual for a member of the cat family, enable the tired hunter to regain its breath while its jaws are still gripping prey. After the exertion of a chase, a cheetah will typically spend 20–30 minutes recovering before starting to feed. During this period, the kill is vulnerable to scavengers such as hyenas and vultures. Cheetahs kill frequently and efficiently, and swallow meat quickly. An individual cheetah can consume 30 pounds (14 kg) in one feed. These characteristics help the animal to compensate for the loss of some of its catches to other meat-eating animals.

Breeding and social organization

Most adult cheetahs live in home ranges of 20–400 square miles (50–1,000 sq km) and mate with individuals from their own range. Apart from meeting to mate, male and female cheetahs do not interact, and the male does not help in raising the young. Cheetah cubs are born blind,

after a 3-month gestation period. Within 4–14 days the cubs open their eyes, and before they are 1 month old they eat their first meat. A female cheetah starts to capture live prey and bring it back for her cubs when they are about 6 months old so that they can practice killing.

The cubs' initial light gray woolly coat is replaced by adult coloration within 4 months. Some cubs remain with their mother for up to 2 years, while others are independent after half this time. Cubs become sexually mature within 2 years. Two or three male siblings will often group together as adults to form a "coalition" and defend a territory, although unrelated males may also do this. Such coalitions usually stay together for life. A shared territory may be as small as 15–30 square miles (40–80 sq km), though most adult cheetahs live in much larger home ranges. Some cheetahs have lived for up to 16 years in captivity.

Long history of population decline

Cheetahs are the oldest of the big cats. Fossils of a cheetahlike animal, *Miracinonyx inexpectatus*, dating back four million years have been found in North America. However, that species disappeared at the end of the last Ice Age. The historic range of today's cheetah extended from India westward to Morocco and southward through the African continent to South Africa. In places, such as some parts of eastern Africa, cheetahs are still relatively abundant, and the species survives in small pockets in some areas of Iran. However, outside Africa cheetahs are either extremely rare or have become extinct. It is very unlikely that any still survive in India.

In 1900 there were an estimated 100,000 cheetahs worldwide; today there are 10,000 to 15,000, 10 percent of which live in captivity. The steep decline of the Indian cheetah was partly due to the trade in coats, rugs and trophies; in 1972 one furrier in New York was found to have almost 2,000 cheetah skins. The spread of agriculture in India robbed the cheetah of its habitat, and its staple food, the blackbuck and axis deer, have been wiped out in many places.

Natural predators have also played their part in the cheetah's demise: 90 percent of all cheetahs in the Serengeti National Park in Tanzania die before they are 3 months old, taken by scavengers such as hyenas, lions, jackals and birds of prey. Cheetahs do not prosper in those parks and game reserves that contain other large predators such as lions and hyenas, because of competition for prey, and many live in unprotected areas, including farmland. Today the killing of cheetahs and trading in their skins is illegal in nearly all African countries and the species is protected by C.I.T.E.S. and the African Convention. However, the future of the cheetah is by no means assured.

Cheetah cubs are born helpless and blind, and the mother frequently moves them to new hiding places during their early life. This reduces the likelihood of the cubs being found and killed by lions or adult male cheetahs.

Index

Page numbers in *italics* refer to picture captions.
Index entries in **bold** refer to guidepost or biome and habitat articles.

Page numbers in *italics* refer to picture captions. Index entries in **bold** refer to guidepost or biome and habitat articles.

Page numbers in *italics* refer to picture captions. Index entries in **bold** refer to guidepost or biome and habitat articles.